Understanding
WATER

Understanding
WATER

Developments from the Work
of Theodor Schwenk

ANDREAS WILKENS

MICHAEL JACOBI

WOLFRAM SCHWENK

Floris
Books

Translated by David Auerbach and Jennifer Greene

First published in German as a special issue of
the journal *Sensibles Wasser* in 1995 by the
Verein für Bewegungsforschung, Herrischried

First published in English in 2002
Revised edition published in 2005 by Floris Books, Edinburgh
This third edition published in 2018

British Library CIP Data available
ISBN 978-178250-506-8

Contents

Song of the Spirits over the Waters

The human soul
Is as water:
From heaven it comes
To heaven it rises,
And down again
To earth it must,
Ever changing.

The pure jet
Streams from the high
Steep cliff.
It sweeps sweetly
In cloudwaves
To the smooth crag,
And, gently taken,
It flows down,
Veiling all,
Gently murmuring
Into the depths.

When rocks jut out against its fall
it foams dissatisfied
step by step
into the abyss.

In the flat bed
It steals its way to valley meadows,
And in the smooth lake
All starry constellations
Delight their countenance.

Wind is the wave's
Sweet lover
Wind mixes foaming billows
From the very bottom.

Human soul
How like water you are!
Human fate
How like wind you are!

J.W. von Goethe

Foreword

Everybody uses water for routine activities, such as drinking, cooking and washing. Therefore all of us should appreciate the idea of relating responsibly to this life-giving substance. In daily use, the 'dirty' water we use has already flowed away before we really notice the effects of our actions. In modern societies it is taken for granted that fresh water will flow to take the place of what has been used. The consequences of our use of water are seldom forced on us. Nonetheless every educated person knows the catastrophic consequences of humankind's ability to destroy. So many plant and animal species have disappeared; so many wells and springs have been shut down because of pollution. The long-term implications of all these effects are still unclear. A few may buy gadgets to protect them from some level of water pollution, but this neither helps solve the main problem nor contributes to a responsible relationship with this most precious element — a substance which so well hides the consequences of our actions upon it, and which is seemingly so plentiful. In the world of inert solids, the remnants of things we have destroyed remain visible to us until we remove them. Sometimes this motivates us to avoid further destruction. This is not the case with water, however, for the used liquid flows away. As the consequences are not forced upon us, learning from our mistakes has to become a conscious activity. We must therefore take the initiative ourselves, and begin to care for water. And in order to learn to care for something it is necessary to come to understand it.

This is the fundamental aim of the Institute of Flow Sciences (Institut für Strömungswissenschaften) at Herrischried, Germany, and to publicize its work in 1994 an exhibition entitled 'Understanding Water' was created, and has been shown in many places since then. The exhibition formed the basis of this book which has since then also been translated into French, Italian and Czech.

The drop picture experiment described in this book is a highly sensitive method which has in the last few years been understood and appreciated to a greater degree. In the 1960s it was used to show the extensive damage caused by detergents, and gained widespread acceptance among the general public, but it was not recognized by professionals as a valid method. As a result of growing interest in chaos research, scientists have recognized that tiny changes have much greater consequences, and have come to respect this approach.

Generally there is a far greater interest today in depicting aspects of water quality. This has also led to questions about the more subtle qualities of water, for instance is technically perfect clean water still healthy, or do the physical purification processes diminish the vitality of the water? This kind of question is an essential part of the drop picture research.

The images in this book are mostly from the work of the Institute, and we would like to thank all colleagues who have helped in the making of this book.

Andreas Wilkens
Institute of Flow Sciences, Herrischried, September 2005

Introduction

We should approach the uniqueness of water with the experience of our senses as well as with our inner experience. A bubbling spring or a babbling brook speaks of freshness and of life. Through the ages humankind has experienced the active forces in water — life-giving, healing, but also disease-generating. Such sensations and experiences allow us to speak of 'living water' although we are dealing with a mineral substance. Water has no biological life of its own, but it enables life the moment it is brought into contact with living matter. Furthermore, without water life is unthinkable. Water is a life-link: watching water awakens us to its mediating role.

Many atmospheric phenomena arise through water in the air as vapour or droplets

The first chapter of the book is dedicated to water phenomena in nature and the different ways in which we interact with them. Then we look at water itself, at vortex phenomena, at water flow and also at unusual aspects which go to make it that vital element of life.

The Institute of Flow Sciences developed new techniques of analysing water, thereby aiming at a deeper understanding of this mediator of life on earth. The drop picture method — developed by Theodor Schwenk in the 1960s for diagnosis and research purposes — has formed the basis of our own research. Our central concern is to understand what good water is and how it changes during its various cycles. This also involves studying the influences of processing and handling of water. The following chapter presents research on the phenomenon of drop-formation and drop-impact onto a water surface. This phenomenon is highly relevant to the drop picture method. Finally there is a description of the Institute itself.

1. Water in Different Realms

When we think of water we might first evoke the reflecting surface of a lake or the babbling of a brook. It is only later that we might think, for example, of clouds. And when we see a round pebble we certainly would not think of water, although the round form is water's work, too.

When we grow aware of how water in our environment appears and what it does, we realize that it reveals itself at various levels, each with its own myriad phenomena. Water in the air, water on the ground: how completely different they are! Then there is that hidden water space below the surface with its own invisible yet so tangible flow. And then there is the ever waxing, ever waning world of form and substance related to water.

These levels speak to us in different tongues. The colour caused by the water in the air and in the light bring forth strong feelings and experiences: we only have to think of the atmosphere at dawn or at dusk. The reflections of a water surface and the waves at play may invigorate us whereas watching and listening to peaceful water-flow may cause us to dream and feel as if we are floating above the ground. Only with the forms which water has created in solids can our object-consciousness really come to grips.

Water cannot be grasped in the same way as other things in the world. It is a master of concealment, withdrawing from so many phenomena, appearing in that which it simultaneously transforms: in the colours of the heavens, for example, or the reflection of a shore on the surface of a lake. Often the more intensely it participates in a particular phenomenon, the more hidden it remains. A moisture-free atmosphere would mean the certain death to all living creatures, a parched earth disintegrating to dust.

Conventional mathematical analysis is hard put to describe even the physics of water, let alone its living and supportive aspects. The formative world of water requires a more holistic understanding of its many levels of reality.

The following pages present a variety of water phenomena on different levels. The kind of activity revealed by the drop picture experiment takes place largely beneath the surface on an underwater level, where various forms of motion occur unseen. It is here that eddies and vortex forms are generated requiring special methods to make them visible.

Water in the realms of air and light

Atmospheric phenomena such as colour in the sky, clouds and rainbows which appeal so strongly to our soul all arise through the presence of water suspended as *droplets* in the air. Falling freely, water forms *drops*.

Mist-droplets   *Freely falling raindrops deformed by air, according to their size (true size pictures)*

Water brightens, darkens and colours the atmosphere

Water on the boundary with air

Still water is like a mirror. In full consciousness we can understand the geometric laws that govern drops, light reflections and wave propagation.

When *excited,* water splashes and forms waves. In a more dream-like state we experience its gentle moving.

In springs water wells up, and in whirlpools it is sucked down. The *welling up* and being *sucked down* show inner motion of the water which we sense almost unconsciously as an elemental world of forces.

Rising bubbles generate ringed waves (springs at Clitumno, Italy)

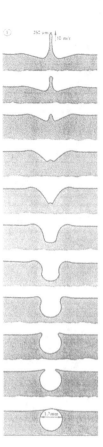

A rising air bubble scatters water droplets into the air. The water evaporates, and salt and organic substances enter the atmosphere. ➤

Capillary waves are an expression of the forces on the surface:

▼ *Pure water: Many capillary waves*   ▼ *Detergent in water: Few capillary waves*

▲ *Capillary waves in a Black Forest brook*

Waves:　　　▼ *(a) By wind*

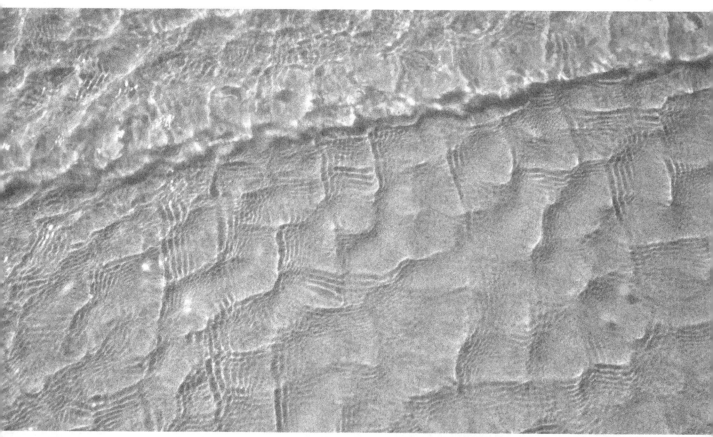

▲ *(b) Through falling water*

▼ *(c) Through flow*

Water driven through the air generates drops and spray

Pushed up or sucked down

Spring-flow and vortex-funnels are witnesses to inner motion: If the surface of the flow is pushed up or sucked down we sense an elemental world of forces with a dull consciousness.

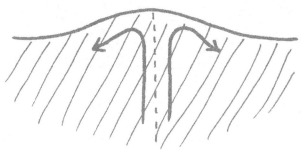

Spring-flow:
Water springs to the surface and raises it

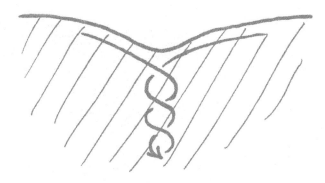

Vortex-funnel:
A vortex, on the other hand, causes a dimple on the surface

Water rotates in a vessel and flows down and out. The water surface has sunk to take on the form of a vortex funnel. ➤

Three moments during the generation of a
vortex ring created by dropping a
milk-drop into water

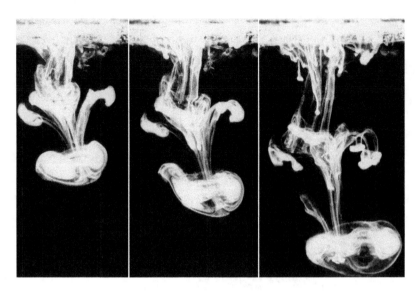

Water beneath the surface

The motion which unfolds itself freely within water generally remains hidden unless it is revealed by adding dye.

A mass of water acts on our body with an inertial force. When we immerse ourselves in water, we lose the sense of our earthly heaviness — the directions of space lose their character, becoming the same, and temperatures equalize.

Looking into this invisible moving world of formation and dissolution we are reminded of the infinitely creative patterns of organic forms.

Vortices are surely the most characteristic expressions of flow in water. When water is stimulated to flow, independent of gravity, a vortex is the response

Vortex flow in a mud puddle

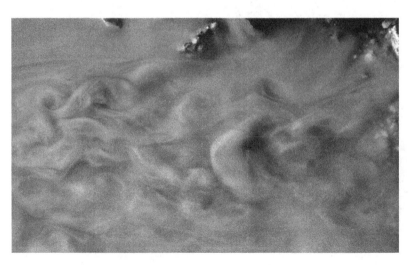

A limestone heap formed by erosion

Water on the boundary with solids

The formative activity of flowing water leaves traces on solids in infinitely varied patterns. In grasping the solidified pattern of the motion that created it, we become fully awake.

Water both creates forms and transforms by dissolving, washing away and depositing solids.

Water is also present and active in solids. It holds substances together, as well as for instance, making them soft and pliable.

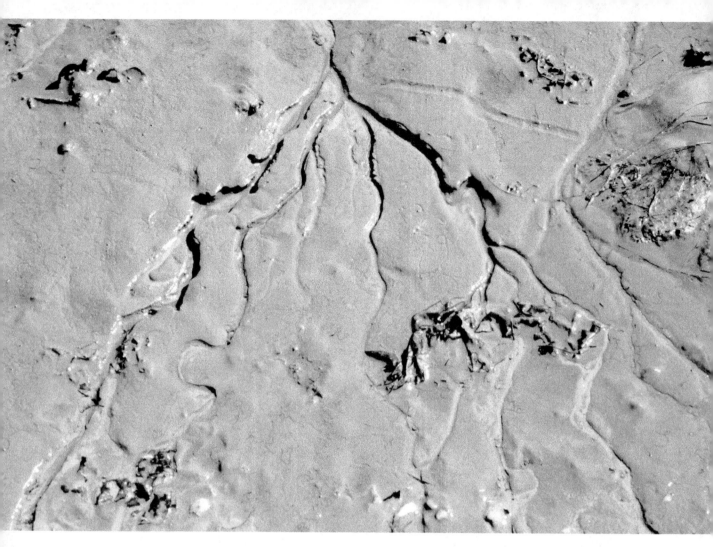

Meander in sand and silt on the North Sea. Water both forms and transforms by dissolving, washing away and depositing solids.

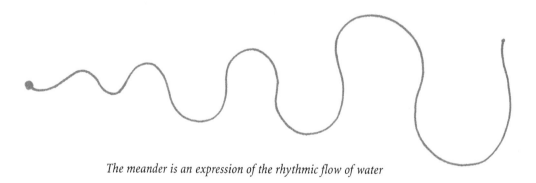

The meander is an expression of the rhythmic flow of water

Meander in stone (Ireland) ➤

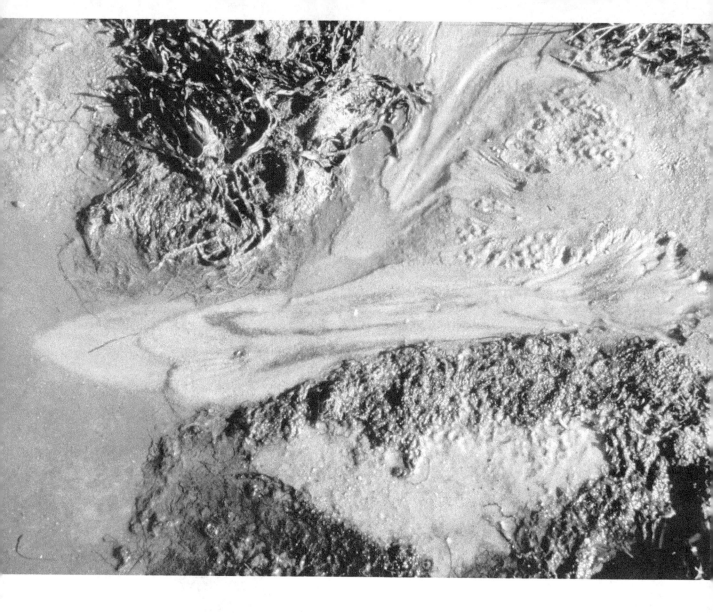

silting
delta ⟵ transporting
meander ⟵ dissolving,
erosion 'tree'

2. Water Phenomena

When we strike a billiard ball it is easy to imagine the linear path it will follow, once we know the laws of its motion. This is the trick of playing ball games. But with water this is not the case! If given a linear motion water reacts by generating rich and varied flow forms, which are bewildering when seen from the point of view of solid-state mechanics. The world of water is one of motion, of becoming and of dissolution, of process. It is therefore impossible to describe the variety of manifestations of water with static concepts. We need to find adequate concepts in order to do this.

Water naturally describes spiralling, sinuous patterns in space from the smallest to the largest scale — patterns which can be recognized as metamorphoses of the vortex form. We may thus see the vortex as one of the most elementary flow forms. However, the concept of form must be understood differently here as opposed to the way it is used in the world of solid bodies. If dyed water flows into clear water and becomes vortical, (forming vortices, see Glossary), we see how forms generate on the boundaries of the dye. The dye boundaries, however, are not real boundaries of the flow, for flow is an unbounded continuum of the entire water mass. The only boundaries are those provided by the water surface and the solid boundary which contains it — the floor of the ocean, the glass-wall of the container, and so on.

In trying to describe the growth and decay of an eddy or vortex, one is drawn to use concepts from the realm of life. If a vortex ring flows against an obstacle it completes a complicated transformation — a metamorphosis. Such transformations, which can go on and on, become possible by the continually changing water mass. This process becomes possible because of the continuous shearing and vortical motions. These motions are essential for the concept of fluids. Vortices generated by shear flow take the introduced dye and draw out the surface which is then wound up in the vortex. All remaining unmixed dissolved matter is drawn out into these surface forms. This process follows laws which manifest themselves in the ring vortex in a particularly beautiful way, reminding us of the development of living organs. We observe how motions meet and lead to the generation of new forms, to their differentiation and finally, to their dissolution when the fluid comes to rest. Quiescence and motion, vortex generation, metamorphosis, decay and again quiescence. The vortex is the mediator between the polarity of motion and stillness.

We thus come to understand water as a substance which both differentiates as well as equalizes, a substance always in an active state of process. Absolutely still water does not really exist: even in a standing glass, water continually evaporates, the water cooling at the surface tends to sink — everything is in flow.

Growth and dissolution, formation and transformation: these are the basics for all biological life. Yet life requires individuation, too: it requires boundaries. This means surface generation right down to the fine structure of the cell. Water allows even this. We thus come to recognize water as the most vital element of life.

In a lecture, Theodor Schwenk said: 'Water does not have the characteristics of the living, but without water there is no life ... Water does not have the ... expressions of life [growth, fertilization, reproduction, metamorphosis, nourishment, metabolism, and so on], but these all only become possible through water ... What is it that enables water to accomplish this? By renouncing every self-quality it becomes the creative substance for the generation of all forms. By renouncing every life of its own it becomes the primal substance for all life. By renouncing every fixed substance it becomes the carrier of all substance transformation. By renouncing every rhythm of its own it becomes the carrier of each and every rhythm.'

Schwenk went on to say, water, 'has no organs, it is the common primal organ of all that lives, of all that remains a functional organ. It does not have the factors of life superficially tacked onto it; it incorporates them all together, but only functionally, as possibility, ability, activity.'[1]

Water phenomena challenge our thinking

In perfectly clear water, flow is invisible

Can we ever really understand flowing water? Being transparent it is invisible. Light and shadow are prerequisites for visibility, so that

▲ *Flowing water in a glass*

► *Pieces of paper floating on the surface make visible the waterflow in a rectangular bath*

the investigator must darken areas of water using particles, dyes or glycerine, for example. Each of these methods in isolation yields an aspect of the motion — a path, a moving point, a streak, a dyed region. One strives in vain to put all these aspects together. We can only begin to grasp the process when our own thinking becomes less rigid and more fluid.

Flowing water generates a whole, but not as a closed system
No matter how isolated the action that causes flow, water responds with a harmonic interconnected response. This does not stop at any boundary — as, for example, the water surface — which would allow one to speak of a really closed system. Every interconnected motion has effects in unbounded space, and can raise our awareness towards the world as a totality.

Water often flows very differently to what we imagine
Even the simple motion of a rod dragged through a shallow water trough has surprising effects. How dismally we fail sometimes in predicting motion in detail! Our way of thinking has to become more

flexible, and image must flow into image, each different, each new, yet all linked. Our thought is challenged to take on the qualities of the phenomenon.

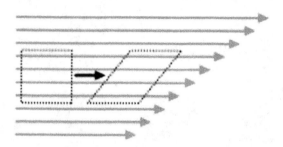

Deformation of an imagined square in a shear flow

Water phenomena in motion demand a transformation within us
Even as we try mentally to grasp the processes of motion, our image-generating mind clings to solid images. Such a way of seeing, trained through observing solids, may be appropriate for thinking about solid bodies. But the same tendency makes us create images of motion in water that are constructed from ideas of solids.

In the realm of fluids, this tendency leads to great difficulties in comprehension. For example, why does water flowing through a constriction generate suction (lessening of pressure) rather than an increase in pressure? Investigating such questions in the world of fluids requires an appropriate thinking, a kind of 'fluid imagination.' It requires mobile pictures. Water requires us to transform consciously our own thinking.

The 'movement response' of water: Variety within unity

— Water reacts to a linear stimulus with rhythmic, differentiated, eddying motion.
— All regions of a fluid are interconnected, harmonically taking part in the motion.

The impression of a totality arises.

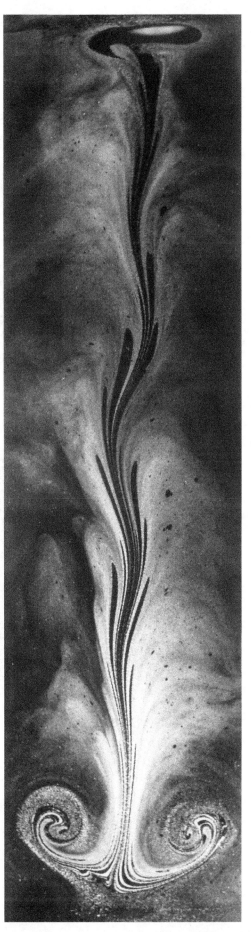

If we place some resistance in the path of the flow, it reacts by moving rhythmically to and fro. We find the most developed pattern formation when it moves not too fast and not too slow ➤

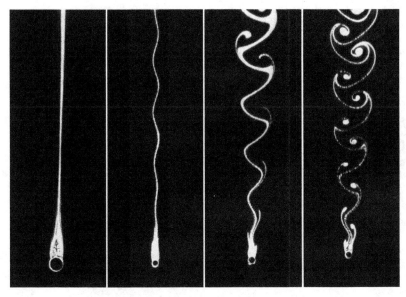

Fluid flows about a still rod (from bottom of picture to top):

— *first slowly: it begins to oscillate*
— *then quicker: meanders and vortices generate clearer forms*
— *if the speed is further increased, more chaotic forms regenerate (not shown)*

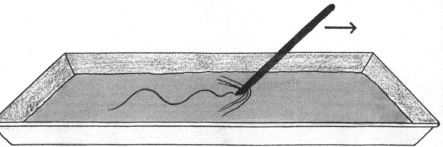

A rod is pulled linearly through water

The flow motion is made visible by means of dusting a powder (or pollen) on the water surface. For better observation the water is mixed with two-thirds glycerine which slows down the flow and makes the forms clearer (see next page).

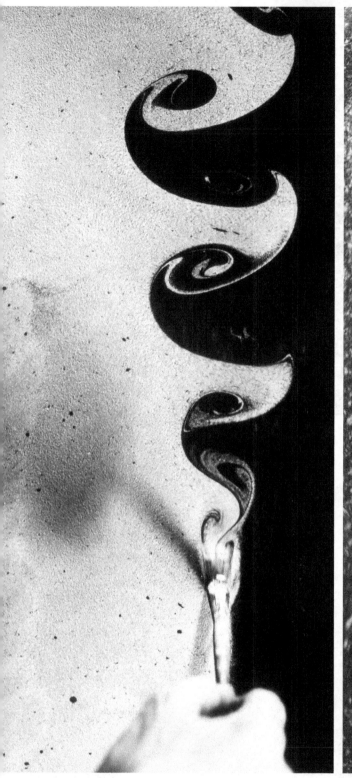

Behind the rod (moving from top to bottom of the picture) the flow gently begins to oscillate

A long photographic exposure enables us to observe the circulating motion

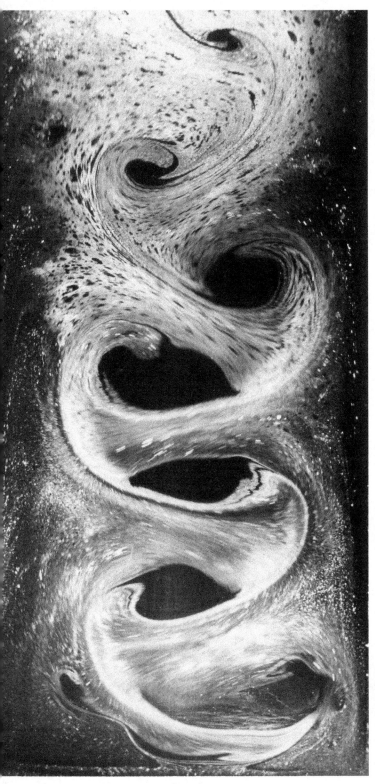

Developing forms become visible with even longer exposure

At rest: the outcome of the formative activity of water

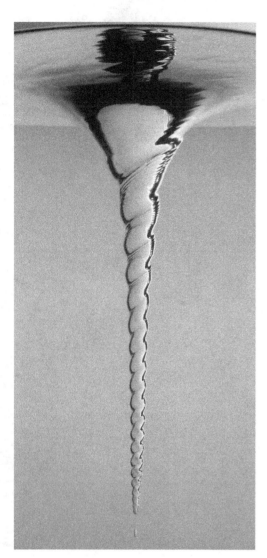

The vortex

The vortex is a flow form. In turbulent flows in nature we find vortices from the largest down to the smallest scale — in ocean flow they may be so large that an entire continent would fit into one.

The circling motion of a vortex always relates to a centre

A vortex forms above the turning water of the outflow, sinking the surface in corkscrew-like tube

From top to bottom of picture ➤

— *Dye is fed into a circulating vortex flow*
— *towards the centre the vortex turns faster. The shearing motion draws the dye into a thin extensive veil.*
— *The surface increases with time*

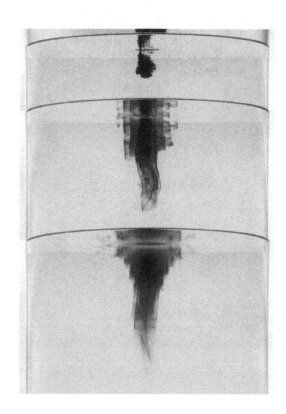

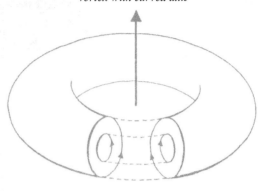

◄ *Water is pressed out of a tube and generates a rising ring vortex, seen here from the side. The centre of its inner circulating motion is ring-shaped. The ring-plane is perpendicular to the plane of the figure.*

Vortex with curved axis

The ring vortex is perhaps the most autonomous pattern amongst flow forms

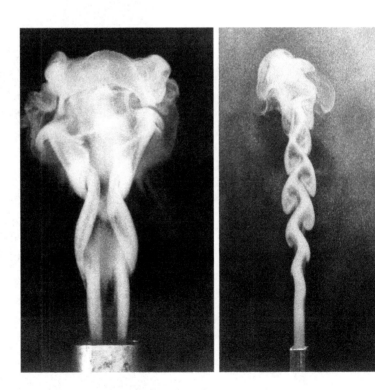

Organ-like forms arise where many vortices interact

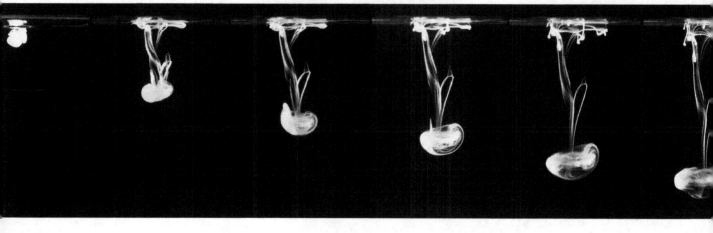

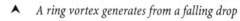

▲ *A ring vortex generates from a falling drop*

◄ *Phases during the vortex generation in the drop picture*

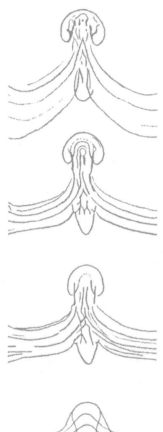

A ring vortex metamorphosis

White-clouded water flows from a tube below into a water container and generates a rising ring vortex.

The ring forms nodes and becomes wavy. When it reaches the surface, it begins turning inside out in complex forms as the veils of dye differentiate.

Thereafter there is motion first towards the centre, then downwards and outwards which, in its turn, generates new vortices.

'I had fewer difficulties in discovering the motions of the heavenly bodies, despite their incredible distances, than I did in investigating the motion of flowing water, something that takes place before our eyes.'

Galileo Galilei (1564–1642)

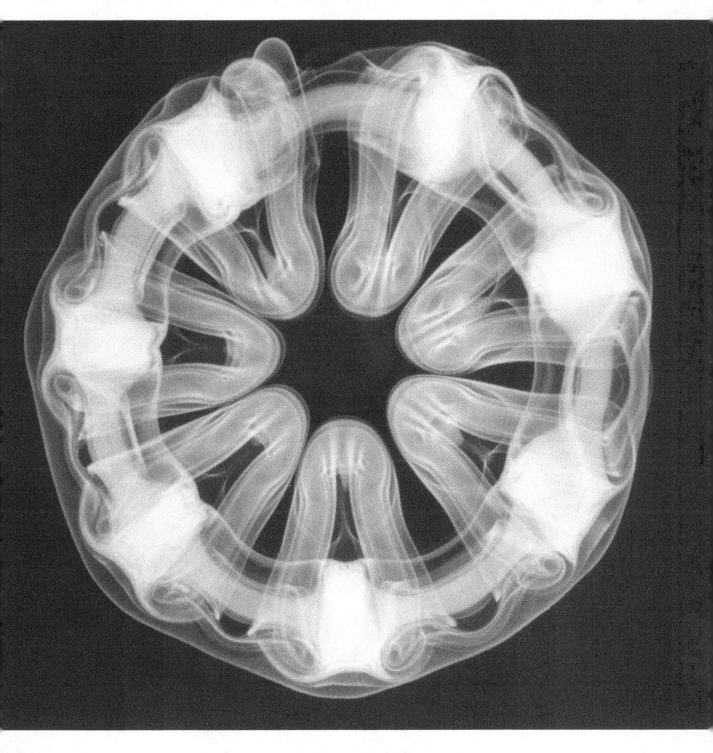

*Above and the following two pages: metamorphosis of a
ring vortex rising towards the surface, viewed from above*

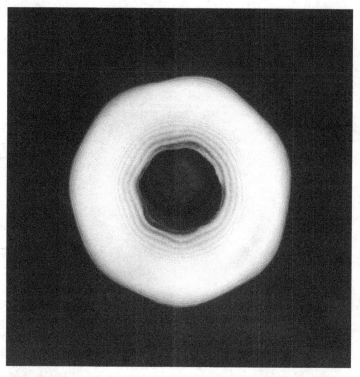
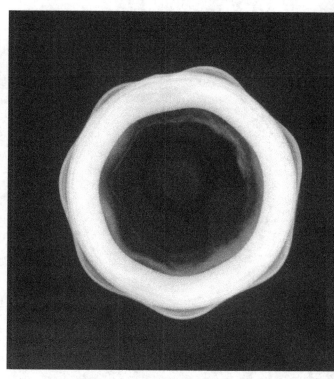
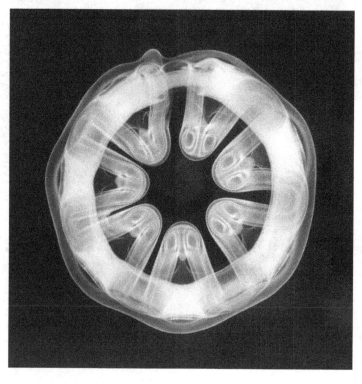
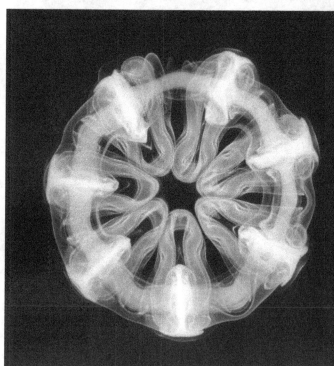

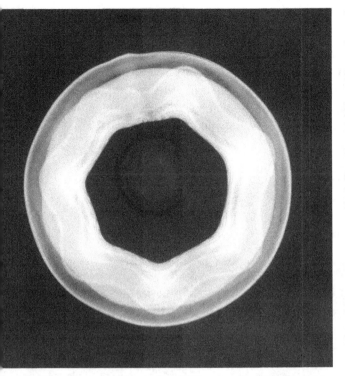

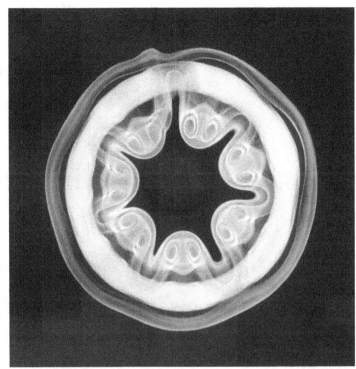

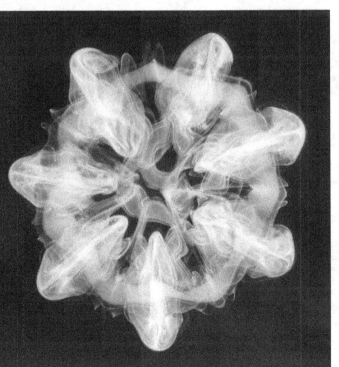

Water movement and water quality

WAVES

Water quality becomes visible in the movement of waves. Two identical pans are filled with equal amounts of water and are simultaneously set in motion. In the right-hand pan one drop of detergent was added to 5 litres of water, equivalent to a dilution of 1 to 250,000.

Left-hand pan is pure water, right-hand pan has a drop of detergent. The surface mirrors the surroundings.

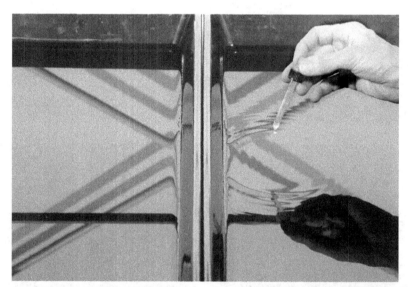

The water in both pans is simultaneously and equally set in motion

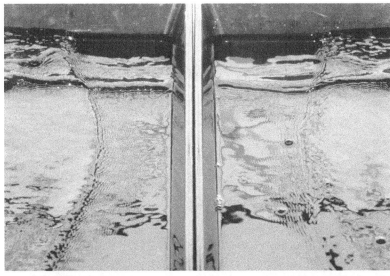

After three seconds the wave motion is visible. It is finer more differentiated on the left.

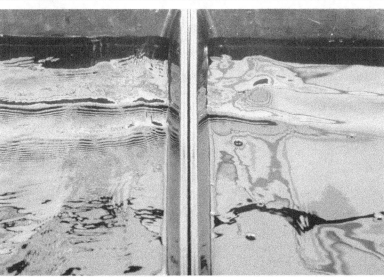

After six seconds more capillary waves are clearly visible in the pure water (left-hand)

After 28 seconds the water with detergent (right-hand) has come to rest again, while the pure water is still in motion, distorting reflections

MEANDERS

Flowing water begins to meander on encountering resistance of solid matter. The meanders show an aspect of the quality of water.

Water flowing down a wet smooth surface begins to meander (ink has been added to show the meander more clearly)

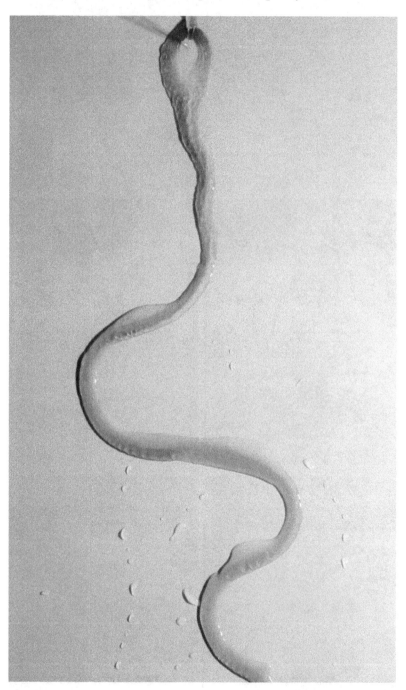

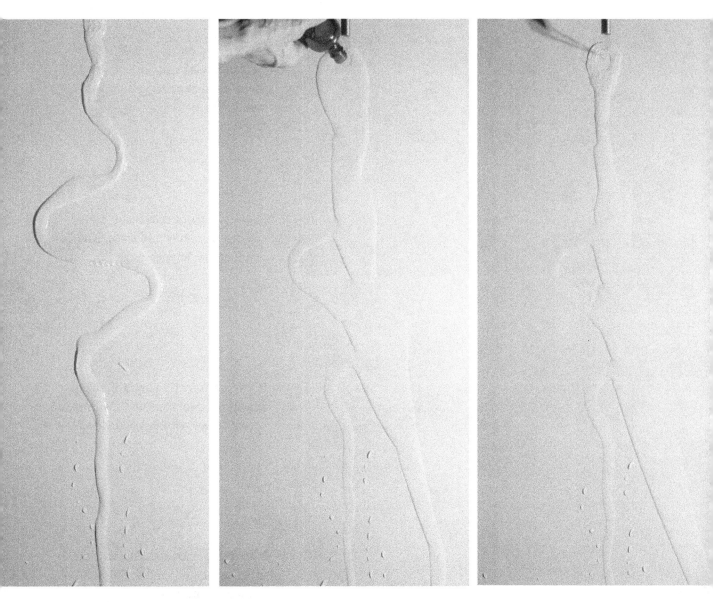

Clean water meanders in tight curves on a varnished surface which is not very moist

A drop of detergent added to the water disturbs the meanders. The water flows in straight, broad lines.

The pure water following, contracts again into a narrower flow, while further down the flow is still broad from the detergent

The polar activities of water

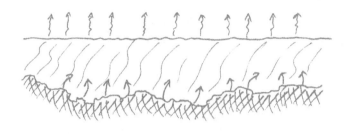

- Becoming one with air and solids in melting: evaporating into air, dissolving solids.

- It levels without destroying barriers: generating an interface and levelling out temperatures.

- It levels out by flowing: following gravity it begins to flow; shearing, it mediates between fast and slow.

- It holds together, forming a whole. It forms a sphere in a drop. So, too, do the world's oceans.

- Flow can differentiate. It begins to circulate in the vortex, creating an individualized but flowing entity. It sets up circulating currents for the earth organism.

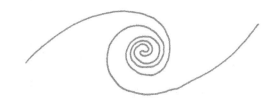

- Differentiation without splintering. In meeting itself, or some resistance, it responds rhythmically, for instance, with a meander.

- When moved out of equilibrium, it responds by forming waves.

- Out of unity into diversity — structuring: condensing, dividing into drops, precipitating.

Water as the basis for life-processes

Water is not just concerned with all life processes, it makes life possible.

Transforming existing structures
Without transformation, what exists cannot adopt a new form. Water creates the possibility for renewal and for metamorphosis, as, for example, we saw on pages 34–7 where the ring vortex turned itself inside out.

Generating new forms
The infinitely creative potential which water presents in its flowing forms and patterns, generates a free space for living beings to take on their own true forms with the help of water.

Maintaining processes
Living beings have no immutable forms like stones. While their forms are maintained, their bodily substances are constantly being renewed. Water, too, demonstrates this quality in, for example, a stationary wave (capillary waves, for instance pp. 15f), a water jet or a vortex funnel (p. 32). These forms remain stable, although the water within the form is continually changing.

Bringing rhythm and structure
The main distinguishing feature of living beings is that they live through rhythms. Plant life develops through the seasonal rhythms of the year. Rhythms of metabolism, breathing and heartbeat animate both humans and animals. When we set water in motion, it begins to move and immediately brings rhythm to the motion.

Moderation
Life on the earth is generally rendered possible by the moderate conditions which water creates. In the atmosphere, for example, water not only keeps off the deadly heat of the sun, but also retains heat at night so that the earth does not cool off too much.

3. The Drop Picture Method

Many people feel that water quality cannot be investigated simply by analysing critical values of undesirable substances present in the water. When one asks the questions: 'What is good water? And how do we distinguish water which refreshes us from water that does not?' one generally meets with uncertainty and vague ideas. There are no firm guidelines here.

In the 1960s Theodor Schwenk, founder of the Institute of Flow Sciences, developed the drop picture method in order to depict water quality in a visual image. The method has now been practised, taught and further developed at the Institute for over forty years. In the standard method, a stream of water is impelled to the sensitive transition range that lies between laminar and turbulent flow. Flow forms generated in this manner express the image-forming qualities of water, in a way that needs to be 'read.' The method enjoys a growing interest amongst experts looking for a more illustrative way of understanding water than that provided by substance analysis alone.

The new aspect — and the advantage of this method — is that one looks at the phenomenon of water on levels other than those provided by its material composition. Using this method one observes the mobility of the water as an expression of its central purpose: namely, to serve life by facilitating (among other things) metamorphosis, growth and decay. There are a number of aims which can be pursued using the drop picture method, but a central theme behind our efforts is to develop a scientifically based evaluation of water in such a way as to arrive at a responsible relationship to it.

Theodor Schwenk observed nature closely and saw how water motion directly revealed qualitative elements. Imagine, for example, the subduing effect on an oscillating capillary wave resulting from the reduced surface tension of a polluted stream. This was one starting point for seeing the mobility of water as an essential expression of water quality in relation to the living world. And this is the basis of the drop picture method. The drop picture method serves to illustrate this mobility. The moving forms under water are diverse complicated transformations of vortex forms which generally remain invisible

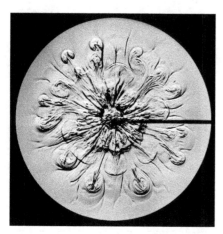

Good spring water

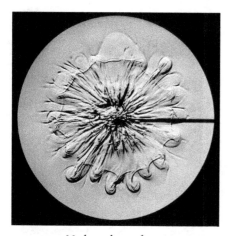

Moderately good water

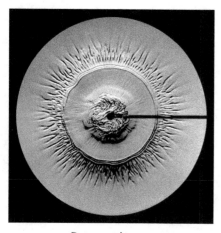

Detergent in water

unless they carry dye or neutrally buoyant particles. In drop picture experiments these forms have to be rendered visible using the appropriate methods. Perhaps the most suited is the schlieren optical method, which uses a special apparatus to detect shimmering effects in the substance (see pp. 52f).

The drop picture experiment

The water sample to be investigated is placed in a special glass dish as a shallow layer. Glycerine is added to the sample which dampens the flow because the glycerine is more viscous than water. Using falling drops of distilled water, the water sample is brought into motion. The glycerine, once mixed with the impinging drop, shows the schlieren. Schlieren appear when, for instance, we see air flickering over a hot surface or we observe sugar being dissolved in water or, as in our case, glycerine being mixed inhomogeneously with water. In all these cases the density of the fluid is not even, but shows density gradients or boundaries which can be photographed with the schlieren optical method.

The actual experiment begins with the impact of the first drop into the sample. But before that can occur a number of preparatory steps have to be taken. Amongst other things, these consist of preparing the laboratory and the glassware; of carrying out a test run with the experimental dish and, an hour before the experiment, drying it, as well as mixing the sample with the glycerine. All steps are strictly standardized.

After the first drop has fallen, a turbulent zone appears in the centre. This is known as the *kernel,* about which a faint and barely differentiated circular line appears as the propagation boundary of the drop.

Using a sample of pure water, one notices as early as the second or third drop how faintly visible vortices shoot out of the propagation boundary, only to begin fading. The subsequent drops amplify this picture until, around the tenth to the fourteenth drop, the kernel begins to transform and lose its sharp contours.

Thereafter new dendrites (tree-like forms) appear at the periphery of the picture which, partly curved, partly branched, grow from drop to drop. We can thus distinguish a turbulent kernel in the centre, a vortex zone about it, which goes through the strongest formation and transformation and finally, an edge region with slowly developing dendrites. Every individual drop picture shows characteristic forms, depending on the water under investigation. Even more revealing is the development of the image during the series from picture to picture. On the one hand, there are experiments with diverse vortex forms which change from picture to picture and are called *multiformed developments*. These experiments follow an unpredictable course and are also called *non-determined*. On the other hand, there are experiments consisting of a single form type during the entire experiment, known as *predictable* or *determined experiments.*

There is no standing still in what is observed here: the observer participates in the continual generation of form, transformation and dying away. This can lead the observer to an excited and enlivening experience. The camera records only one climax of this development. In order to concentrate the investigator's attention, observations are constantly recorded in a special kind of shorthand.

The development, experienced properly, reveals in picture form the true and meaningful character of water's mobility and openness. Using the image, and working with a progressively greater understanding of the characteristics required for good quality water, the investigator arrives at the ability to make a qualitative assessment.

Various patterns of the flow form on being stimulated in the same way

Drop detaching from a needle

Needle with drop and drop picture dish in the apparatus

The basic drop picture method

The drop picture method helps visualize the tendency of moving water to create forms under standardized conditions.

1. A new drop of distilled water forms on the tip of a specially prepared injection needle every five seconds at a height of 10 cm above the sample.

2. The separation from the needle makes the drop oscillate.

3. The sample, mixed with 13 percent of glycerine (by mass), is placed to a depth of 1.1 mm in a flat glass dish with an inner diameter of 14 cm.

4. The impact of the drop on to the sample makes a crater which forms waves (see p. 91).

5. The drop picture dish stands horizontally in a schlieren apparatus, which helps to make visible the schlieren (shimmering effect) caused by unevenness in density in the water/glycerine mixture.

6. The schlieren become visible, as if in relief, as alternating light/dark structures. In the layer beneath the surface new flow forms are generated after every drop impact. These partly dissipate (vortices, rosettes, garlands) while other structures (lines, rays, dendrites) are also generated which grow more slowly and linger.

During the impact of the drop onto the water layer, the flow process passes through an unstable phase between steady laminar flow and eddying turbulence. This leads to more or less sharp patterns in the layer. Due to the openness of these unstable flow processes, the generation of a drop picture is sensitive to the smallest differences in the initial conditions. Thus, different water samples lead to different forms in the drop picture.

The experimental sequence

The initial conditions in a drop picture experiment are set to be in a sensitive equilibrium. These have to be exactly maintained in order to arrive at meaningful results. The sensitivity and repeatability has been further refined and requires experience, practice and sharp attention on the part of the experimenters.

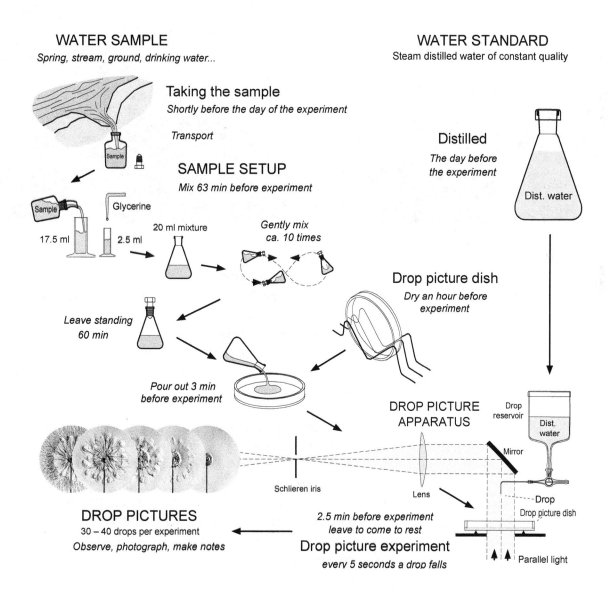

WATER SAMPLE
Spring, stream, ground, drinking water...

WATER STANDARD
Steam distilled water of constant quality

Taking the sample
Shortly before the day of the experiment

Transport

SAMPLE SETUP
Mix 63 min before experiment

Sample

Glycerine

17.5 ml   2.5 ml   20 ml mixture   *Gently mix ca. 10 times*

Distilled
The day before the experiment

Dist. water

Leave standing 60 min

Drop picture dish
Dry an hour before experiment

Pour out 3 min before experiment

DROP PICTURE APPARATUS

Drop reservoir

Dist. water

Mirror

Schlieren iris

Lens

Drop

Drop picture dish

DROP PICTURES
30 – 40 drops per experiment
Observe, photograph, make notes

2.5 min before experiment leave to come to rest

Drop picture experiment
every 5 seconds a drop falls

Parallel light

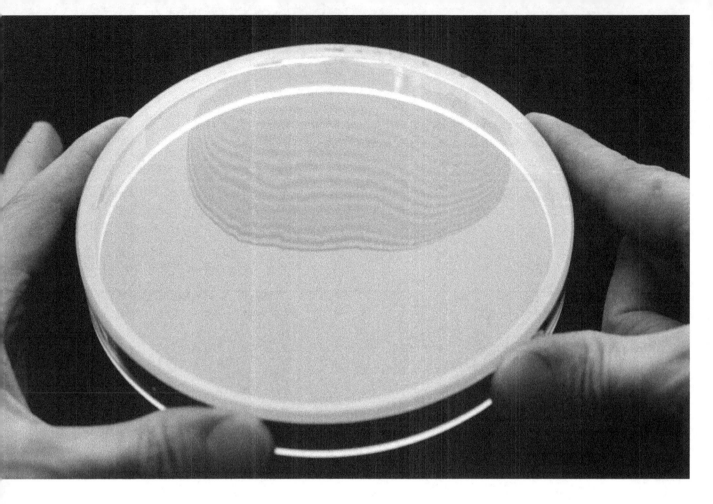

Drying drop picture dish as seen from below

Experimenter and observer

Without an experienced and practised experimenter who has mastered the procedure, one cannot achieve reproducible experiments and results. Implementing the drop picture method requires detailed introductory training of at least three weeks of whole day sessions. Setting up and running a drop picture laboratory only pays — out of reliability and cost considerations — if it is continually used.

There are many experimental prerequisites to ensure the success of the drop picture method.

Cleanliness of the glass

Even slight amounts of dirt or traces of detergent, can influence the experiments. Therefore, cleaning the apparatus is of special importance. This is done using a special method developed for this purpose.

THE DROP PICTURE DISH

After good wetting when the dish is held askew to dry, a thin water-film develops. This appears in rainbow-like colours (Newton's rings). This phenomenon makes the slightest perturbations on the glass surface visible. Additional control experiments serve to assess the correct condition of the drop picture dish.

WATER STANDARD

The drop-water and the sample-water for a control experiment — called 'water control' — serve as a comparison with other samples. It consists of distilled water of a consistent quality as possible and is carried out using water from one of our own granitic rock springs.

THE GLYCERINE

The glycerine, which is mixed with the sample under investigation must be absolutely pure and of a consistent quality, so that it does not change the sample in an undefined manner.

DROP SIZE

A drop weighs 15.53 mg and impinges onto the sample fluid from a height designed to let it strike at a specific phase during its oscillation.

THE SAMPLE

The water sample must be gathered according to instructions in specially inspected bottles and, in the case of sensitive samples, should be tested soon after gathering.

The drop picture schlieren apparatus

The drop picture apparatus consists of a flow-generating source and a visualization device. The flow is generated by the impact of a falling drop on to water in a glass dish, and the image is captured by Töpler's schlieren optical method. The flow forms become visible with the help of glycerine in the sample.

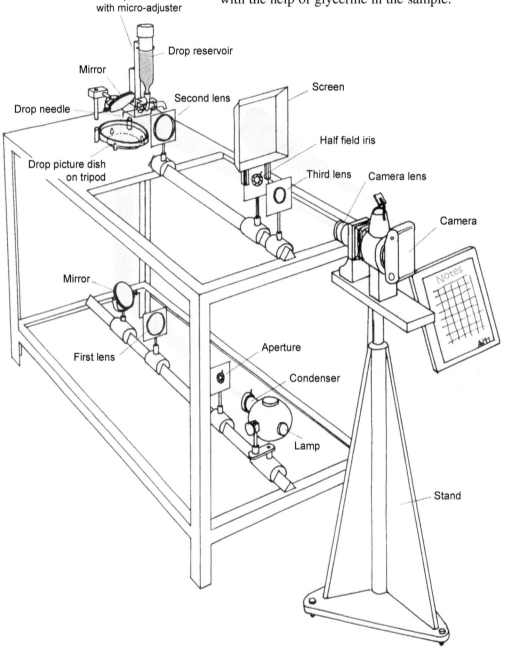

Drop reservoir with micro-adjuster

Drop reservoir

Mirror

Drop needle

Second lens

Screen

Drop picture dish on tripod

Half field iris

Third lens

Camera lens

Camera

Mirror

First lens

Aperture

Condenser

Lamp

Notes

Stand

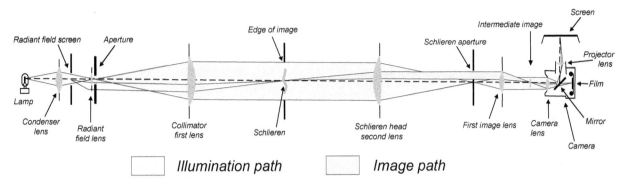

Illumination path   Image path

▲ *Illumination and image paths of the schlieren apparatus*

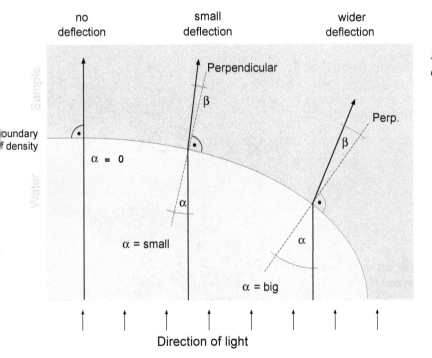

Deflection (or refraction) of light passing a boundary of density at different angles

A drop picture standard experiment

The standard experiment is with pure water. A selection of magnified samples between the first and the 47th drop is shown on pages 54–7.

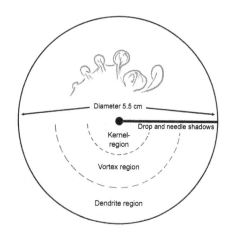

Diameter 5.5 cm

Drop and needle shadows

Kernel-region

Vortex region

Dendrite region

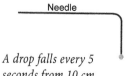

Needle

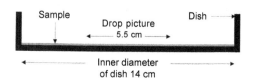

Drop
Distilled water

A drop falls every 5 seconds from 10 cm high into a 1.1 mm thin layer sample mixed with 13% glycerine

Sample

Drop picture
5.5 cm

Dish

Inner diameter
of dish 14 cm

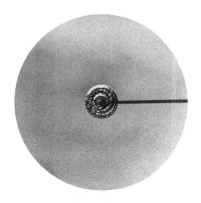

Two seconds after the first drop of distilled water plunges it has spread to a barely visible veil

First drop

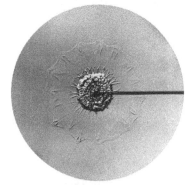

The first vortex rudiments become visible

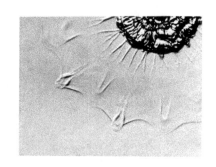

Second drop

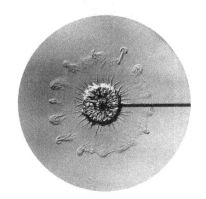

The picture becomes more differentiated with each further drop

Third drop

54

Fourth drop

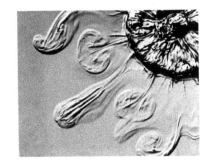

Vortex heads with stems: primary vortices

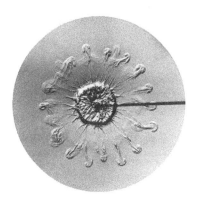

Fifth drop

Broad vortices beginning at the kernel and always alternating with primary vortices: intermediate vortices

Eighth drop

The spreading front has generated only the rudimentary vortex form: garlands

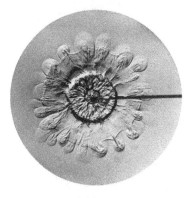

Eleventh drop

The kernel region is bounded by a sharp edge: distinct kernel

55

The kernel begins to fade, the boundaries are no longer as sharp: soft kernel

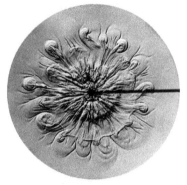

Line structures begin to generate and branch: dendrites

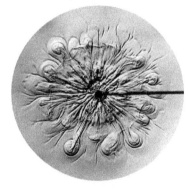

Vortical type

A new picture forms from drop to drop. Regions with strong vortex development may have few vortices in the next picture and vice versa: multiformed development

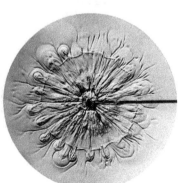

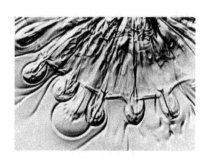

Garlanded type

25th
drop

Kernel and vortex zones are interspersed with line-forms: dendrites

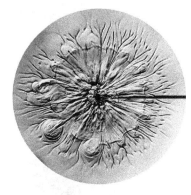

30th
drop

The boundary of the kernel has vanished

40th
drop

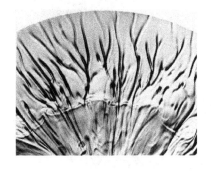

The dendrite region is fully developed

47th
drop

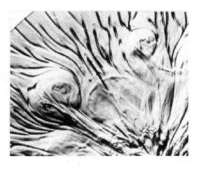

Between the dominating line-structures, the still powerfully developed vortices become ever less clear and begin to fade more quickly

57

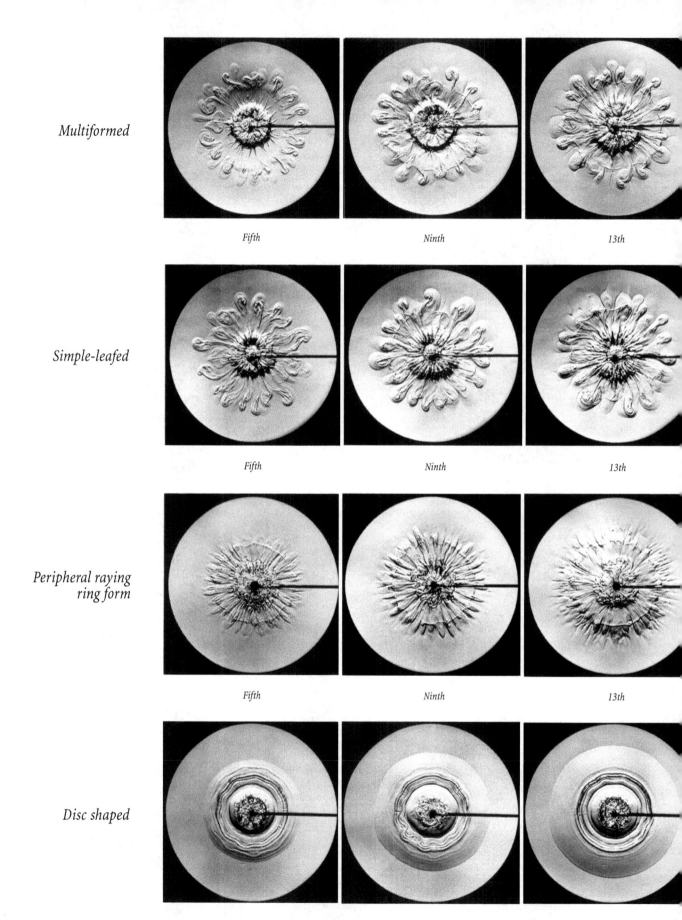

Multiformed

Fifth   *Ninth*   *13th*

Simple-leafed

Fifth   *Ninth*   *13th*

Peripheral raying
ring form

Fifth   *Ninth*   *13th*

Disc shaped

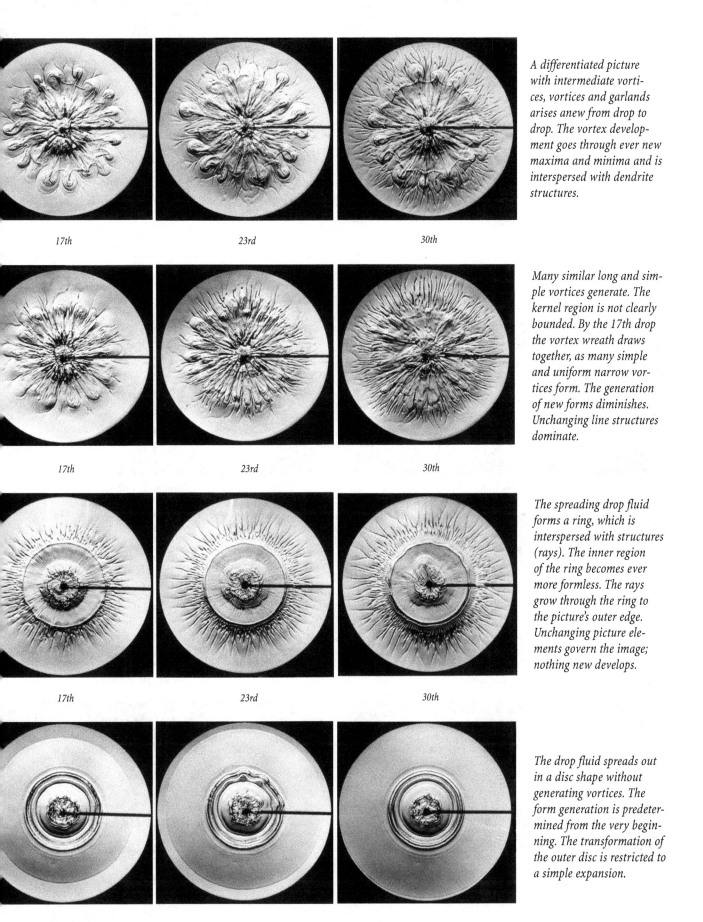

A differentiated picture with intermediate vortices, vortices and garlands arises anew from drop to drop. The vortex development goes through ever new maxima and minima and is interspersed with dendrite structures.

17th   23rd   30th

Many similar long and simple vortices generate. The kernel region is not clearly bounded. By the 17th drop the vortex wreath draws together, as many simple and uniform narrow vortices form. The generation of new forms diminishes. Unchanging line structures dominate.

17th   23rd   30th

The spreading drop fluid forms a ring, which is interspersed with structures (rays). The inner region of the ring becomes ever more formless. The rays grow through the ring to the picture's outer edge. Unchanging picture elements govern the image; nothing new develops.

17th   23rd   30th

The drop fluid spreads out in a disc shape without generating vortices. The form generation is predetermined from the very beginning. The transformation of the outer disc is restricted to a simple expansion.

Development types in the drop picture experiment

Pictures, pp. 58f

It is not the single picture but rather the transformation of the pictures during an experiment which yields information as to the water sample's ability for renewing; its openness, its restraint and predictability of shape formation.

Various drop heights

Showing the ninth and 20th drop picture

Standard height 10.15 cm

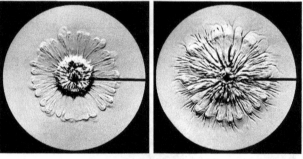

Height 10.5 cm

Height: 10.8 cm
(height = needle exit to dish floor)

Various layer depths of the sample

Showing the fifth and seventeenth drop picture

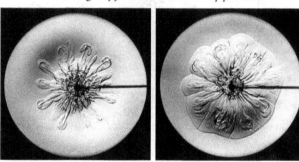

Twice the sample volume

Standard sample volume

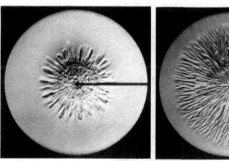

Half the sample volume
(all experiments with the standard glycerine concentration)

Drop impulse and layer thickness

It is the drop that induces the flow process. During the fall it oscillates (see p. 82). Each height yields a different drop form which leads to different sample patterns.

At a height of 10.15 cm the lower side of the oscillating drop is not strongly deformed: multiformed experiments result. At 10.5 cm the

Viscosity:

Showing the fifth and twentieth drop pictures

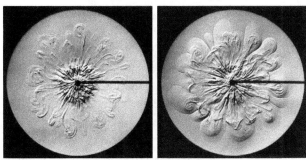

Half the glycerine, lower viscosity

Standard glycerine/water ratio

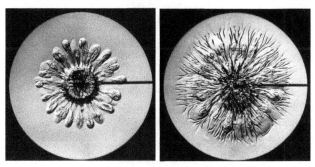

Twice the glycerine, greater viscosity

Density:

Showing the fifth and twentieth drop pictures

Lower density in pure water

Increased density by adding a 3% dissolved salt

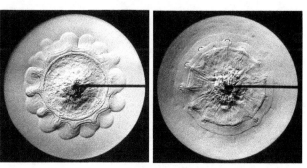

Strongly increased density, 10% dissolved salt

drop is somewhat flatter: a daisylike vortex wreath evolves throughout the entire experiment. At 10.8 cm the drop is more curved: The kernel region and the vortices become misshapen, large and fat. Many rhythmical rosettes evolve throughout the entire experiment.

The sample fluid layer creates a free space in which the flow forms can unfold. In the deeper sample, the drops may spread out more in a vertical direction and induce fewer vortices towards the periphery. Under standard conditions there is maximum spread of the vortices to the periphery, and this spread is well-balanced with dendrite formation. In the shallow sample there is no space for the vortices. The slow distribution of the drop fluid at the periphery permits raylike forms to generate.

Density and viscosity: fluid dynamic parameters

The physical interrelationships — not the actual substances — reveal themselves in the flow picture. When substances dissolve in water they generally also change their physical qualities. Both the density as well as the viscosity have a significant influence on the flow, as shown in the examples of the denser salt and the more viscous glycerine.

On halving the amount of glycerine the vortices shoot out without resistance. Yet with double the glycerine they are clearly restrained and cannot develop. As the density of the sample solution increases, tangential form elements, which can no longer be dispersed, appear increasingly. We call such pictures 'swollen.'

In order to obtain drop pictures with the maximum predictive power and sensitivity, an optimal relationship of the physical conditions for the experimental set-up must be set and maintained.

4. Research Results
Using the Drop Picture Method

Relation to dissolved substances

The drop picture flows are influenced by substances dissolved in the water. This is due to their concentration and subsequent flow-related qualities rather than to their chemical properties. These qualities include mainly the viscosity, density and surface tension and are a holistic expression of their interaction. Therefore, drop pictures can make neither chemical nor bacteriological analytical statements. Similarly, the influence of minerals in small concentrations is minimal. Higher concentrations with their corresponding higher density and viscosity, become noticeable in the 'squashed' forms in drop pictures. On the other hand, organic substances may influence the flow more significantly and shift its unstable equilibrium. Next to the viscosity and the density, the surface activity calls forth an extremely sensitive reaction in the drop pictures.

DRINKING WATER

Numerous experiments with drinking water using the drop picture method have shown that unpolluted, pure natural groundwater and springwater shows a multiformed response in drop picture experiments (see pp. 71–3). This mobility is generally not attained by hygienically treated tap-water obtained from polluted raw water: The latter flows unrhythmically, lacking forms and is poorly differentiated. Nowadays, in Central Europe one might find simple-leafed series of images, while in the 1960s in such cases one might even have found ray-formed, ring-formed or disc-formed series, showing an improvement in the quality of municipal drinking water in the last forty years. As a complement to analytical findings, the drop picture method can be used to ascertain to what extent a drinking water fulfils the DIN 2000 (German standard for drinking water) guidelines, in terms of the character of its mobility. Standard analysis considers only the absence of unwanted elements, while through observing water's motion behaviour it is possible to describe its positive qualities given as desirable within the guidelines. The formative mobility of water should be seen as an independent aspect of its quality. As a complement to the hygienic-analytic findings, our approach therefore contributes to a holistic concept of quality.

Multi-stage biological processes of water purification — such as seepage and retrieval through woodland soil, through sand or biological filtration processes — can lead to multiformed drop picture runs. Similarly the underground passage of water through dunes shows a higher quality in the forms revealed in drop picture experiments.[3]

The substances of pipelines and containers may also affect the mobility of water. Transport containers and house installations made of synthetics showed an impeding tendency on the quality of forms; this is revealed particularly in the case of hot water systems (see p. 77).

The drop picture method responds particularly sensitively to surfactants like detergents (see pp. 78f). However, the calcium content of otherwise pure water does not affect its mobility in the drop picture method.

SURFACE WATER

One of our projects has involved determining the self-purifying qualities of an eight-km section of stream into which brewery wastewater flowed (see pp. 74–6). Perhaps the most important aspect of this work was that we compared the drop picture method with chemical and biological methods for judging the quality of the water. Before the discharge of wastewater, the stream water showed a multiformed drop picture with a differentiated development during the course of experiments. Drop pictures of the section directly after the discharge of wastewater show pictures with nothing more than concentric rings during the whole series of images. By the end of the section the diversity of the forms has grown to that of water before pollution.

The original unpolluted water shows similar pictures to that after the self-purifying section, despite the fact that its composition is now different. The organic pollutants become mineralized by microorganisms and other small living creatures. They have been reintegrated into natural cycles and can be used for new life. The drop picture is an illustration of this.

If one compares the organisms present during the various phases of the self-purifying section with the drop pictures made there, one becomes aware of morphological parallels. In the zone containing the greatest pollution, we find wormlike structures with a repetitive form element. Here, there are only a small number of species, each with a relatively large population density. The drop pictures show a monotonous series of concentric rings over all the runs of an experiment. During the course of the self-purifying section, the population density sinks while ever more new species appear. These have more differentiated bodies as well as sense activities while having gone through more differentiated development stages. The drop pictures show corresponding qualities: the forms become more diverse and the runs show a more differentiated development.

The drop picture experiments show a similar integrating picture of the instantaneous water composition in the same way as the biological population shows a holistic, though temporally, integrating picture of the water quality. It is here that one moves on to a new level of the natural world, on which the essential character of elements begin to speak their language. This is another level beyond that of substances — which are the building blocks for these phenomena. One learns how water quality possesses various forms in a particular living environment and how it can be judged within this context.

Evaluation and judgment of drop pictures

The drop picture method allows us to view a particular dimension of the natural world. It is a holistic method, but this should not be confused with being all-encompassing. In the drop picture experiment something is expressed which requires a special way of seeing to be developed through intensive practice. In this way evaluation criteria arise: one can find that the mobility of water reveals itself in a pictorial manner and that this is meaningful in life terms.

In order to determine the development of this mobility and to compare it more accurately in quantitative terms, we have developed a drop picture morphology which can be used as a scale for the appearance of vortex structures. This differentiated evaluative method allows the distinction of tiny differences between different samples. Although it is only one way of evaluating amongst many, it is, at the moment, the most important tool for distinguishing relatively pure waters, for instance, drinking waters.

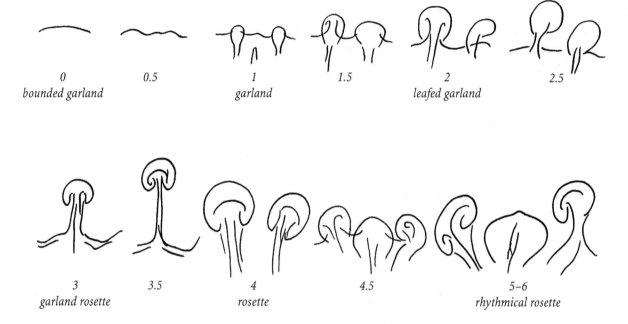

0	*0.5*	*1*	*1.5*	*2*	*2.5*
bounded garland		*garland*		*leafed garland*	

3	*3.5*	*4*	*4.5*	*5–6*
garland rosette		*rosette*		*rhythmical rosette*

Here is an outline of the method. The drawing above illustrates typical vortex forms from drop pictures in a series showing how they develop. Each degree of development is then given a number. When a drop picture experiment is evaluated we first determine the degree of the vortex development of each picture which is then transformed into numbers and can thus be graphically represented. On the chart,

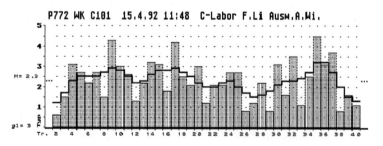

Quantitative evaluation of the standard drop picture experiment of pages 53–7 in graphic form. The rising and falling of the bars shows the change in the degree of vortex development during the experiment. As seen here it can change markedly. The thick line through the bars shows averages of three values.

the vertical axis shows the values for the development (0 to 5) while the horizontal axis shows the drop number (2 to 40). The curve thus generated allows one to recognize when the sample shows a greater or lesser tendency to form a vortical picture. The basis of this drop picture typology has been published in *Sensibles Wasser*, Vol. 2.

COMMENTS ABOUT THE METHOD

We use the drop picture method on the one hand for basic research into the emergence of shape in flows, and on the other hand for a more application-oriented approach where the subtle qualities of water are investigated. A reliable use of this method requires one to know and to take into consideration a number of conditions for which an introductory training by an experienced scientist is vital. The Institute offers introductory courses from time to time.

The value of research using the drop picture method

Since the drop picture method has a very specific role to play amongst many other evaluation methods, we would like to offer a few basic thoughts.

Our relationship to the natural world is largely determined by the viewpoint we construct. The most common human attitude arises out of our self-centred desire to control and use nature to our own ends. Such an attitude, which fails to recognize any needs in the world other than our own, has consequences which rebound on us, as can be seen by current global environmental problems. The serving character of water cannot be understood when questions are asked in this spirit. Asking what water really is and how it interacts with others arises from a completely different starting point. Our very attitude must become selfless! The only realities we can truly understand are those which live in each of us. Therefore scientists themselves must change when trying to understand and associate with water. The drop picture method was developed from anthroposophical ideas and is a step in the above-mentioned way of questioning, a way which we feel is necessary for future development. If one wishes to understand the value of the drop picture method, one must be prepared to adopt different points of view from those which customarily inform conventional research.

Research using the drop picture method may be considered on a hierarchy of levels. On the lowest and simplest level we can distinguish differences between various water samples using drop picture experiments. We can describe the forms they reveal in the drop pictures. We thus can experimentally determine changes that may take place, for example, when water has been treated (assuming, of course, that the method is sensitive to these changes). This holds both for material, as well as for dynamic changes (for instance, by shaking). Here, we have in mind the example of the potentization of homeopathic medicines. The method is very sensitive to changes in organic substances, in particular, to surfactants (surface active agents). Mineral substances are more closely related to water and only become manifest in higher concentrations in drop pictures. If we are interested in the change in water without going into qualitative aspects, we use this method as indicator in the conventional manner, with the exceptional feature, however, that we are not involved with numerical values but rather with pictures. Thus, rather than being a purely analytic method, it is a diagnostic method. Examples of its use in this way may be found on page 77.

At a higher level, we use the method effectively as an organ for observing the mobility of water as it manifests itself through drop pictures, for instance, in the course of a river, in technical processes, and so on. Here the region from which the water sample originates and the picture of the water in the experiment mutually illuminate each other. One learns something about the stream, about the water and about the method. Such experiences allow the method to tell its own story and we learn to see how qualitative aspects in the immediate life-relationships where the water originated, begin to speak their own language. We thus gain insight, not only into the qualities of the mobility of water which express themselves in the drop pictures, but also to the way in which these qualities are embedded in their natural surroundings. If this quality is lost by man-made pollution, for example, then a part of the natural life accompanying this water is also lost and, if any, only lower forms of life establish themselves. An example of this aspect can be found on pages 74f.

We thus acquire criteria for forming assessments and discover a hitherto unknown quality which now gains significance: the mobility of water. A particular value of the drop picture method at this level is that we are able to characterize a good quality drinking water positively by its mobility. A method of evaluation has been developed which allows a critical judgment of this aspect of mobility via the morphology of the vortex forms (p. 66).

If we imagine two drop pictures, each showing the same number of comparable vortex forms in the above sense, they may nonetheless have a different character and make very different impressions on the observer. One might appear ordered, harmonious and forceful, whereas the other appears chaotic, disharmonious and insipid. In order to capture this sort of aspect, we need a level which gives access to the image aspect of the pictures, a step to finding formative processes active in nature. We find a starting point in this direction in the spa water investigations (p. 71). Here we find an image of the physiological function which the water addresses as a healing quality. Such examples are only pilot experiments pointing along the path we are seeking. The significance of this sort of work was described in the following manner by A. Selawry commenting on Ehrenfried Pfeiffer's method of sensitive crystallization:

> This method opens a possibility of bringing the essentials of form to our consciousness today. If we can grasp the activity of a formative world in phenomena, then science will find a bridge from forms to those form-giving ideas or archetypal images. This will open the way to the science of tomorrow.[4]

Mountain spring
multiformed-rhythmical development

Mountain stream
multiformed to simple-leafed development

Wastewater-laden mountain stream
ring to disc-shaped development

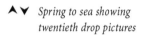 ▲▼ *Spring to sea showing*
twentieth drop pictures

From the spring to the sea

'The springs are really the eyes of the earth. With the ocean the earth does not look into space because the ocean is salty, and this causes it to become only internal in the same way in which our stomachs are internal. The springs, which have fresh water are open to outer spaces and are like our eyes which open themselves to the world around.'[5] Rudolf Steiner

Lower Rhine
Simple leafed to raying development

North Sea
(far off-shore) — water surface

North Sea
(far off-shore) — 15 m deep

Spas

Spa waters are medicines, not foodstuffs. They address the organism in a one-sided fashion. Their flow behaviour is, typical of the individual spring, more or less constrained, balanced or rayed.

Spa waters showing 25th drop

Bad Kreuznach — Elisabethspring

Wiesbaden — Kochwell

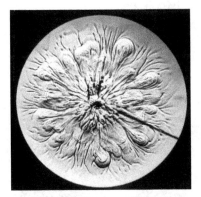

Bad Schwalbach — Weinwell

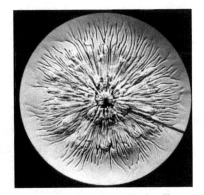

Schlangenbad — Trinkwell

Drinking water investigations

What constitutes good drinking water? Naturally pure and unspoiled groundwater or springwater — the yardstick of good refreshing water — shows a common comprehensive type of flow behavior in drop pictures independent of its calcium content. In a lively flow, manifold rhythmic forms are generated, forms which change and renew themselves with each new impulse. This water, with its mobility, shows a maximum of formative capacity.

On the other hand, water damaged by human activity flows unrhythmically, is poor in formative capacity and is less differentiated. We thus see how the form-generating flow behaviour of water may be seen as an independent aspect of its quality which, as a complement to chemical analytic findings, can make a statement about good, enlivening water.

A particular water is potable when it is hygienically pure. The drinking water law regulates necessary steps to determine this. A drinking water is 'good' in a comprehensive sense when it is hygienically pure and, over and above this, when its other qualities also concur with those set out by the DIN 2000 guidelines. Amongst these qualities are its formative mobility. Such water is experienced as refreshing, thirst-quenching, as 'living.' This is also possible when the water is hard.

Drop picture investigations show the mobility aspect of water rather than the potability aspect; the latter must be tested with chemical and bacteriological analysis. One cannot tell from the flow forms which individual substances are dissolved in the water, but they do yield a picture of how these substances affect the mobility of the

Naturally pure, very soft springwater from granitic rock

Naturally pure, drinking water from medium hard groundwater

Naturally pure, very hard Jura spring water

water, and how the water with all its substances responds to them.

Drinking water which corresponds to the DIN 2000 guidelines is hygienically pure because of its origin. Its mobility shows itself in a richness of forms, coupled with a large variety of differentiated flow patterns. This is barely influenced by the hydrochemical properties, for instance, its hardness.

Drinking water from impure raw water is made hygienically pure in a water treatment plant with technical purifying methods. It does not meet the guidelines of good water. Its drop picture flow is compacted, developing simple, relatively immobile pictures and series.

Self-purifying section of a wastewater-polluted stream

Brewery wastewater enters the Mettma, a Black Forest stream. Through the organic self-purifying ability of the stream this water becomes biologically decomposed during its course of eight km. Organic substances are mineralized by the organisms of the brook.

An unpolluted stream is an environment for manifold organisms. Together, they form a living community of plants and animals: water plants, algae, insects, their larvae and pupae, worms, crustaceans, gastropods, micro-organisms. These organisms are distributed in a balanced manner throughout the environment.

German guidelines for the requirements of good drinking water (DIN 2000):

• *The quality requirements for mains drinking water must in general be oriented on the qualities of unspoilt groundwater of fresh and impeccable constituency extracted from the natural cycle.*

• *Drinking water must be free of pathogens and may not have any health-damaging qualities.*

• *Drinking water should be appetizing and enjoyable. It should be colourless, cool, odourless and of impeccable taste.*

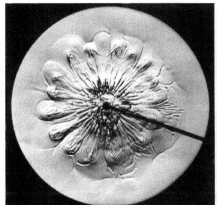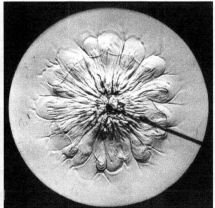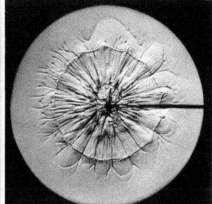

Drinking water from groundwater, which has been mixed with purified river water (from the industrial regions of the Rhine-Ruhr and mid-Elbe rivers)

The Mettma, a naturally pure mountain stream, upstream of the source of pollution

The Mettma at 300 m during flooding

Sampling point upstream of the wastewater discharge. Clear pure water. The drop picture is balanced, multifoliate. A variety of plant and animal communities live in the water.

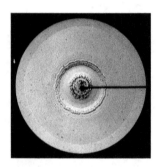

Brewery and domestic wastewater were discharged into the Brünlisbach until 1977

After wastewater discharge, the riverbed is covered with a grey bacteria coating in which colonies of tubificides dwell.

(c) Jointed worm

(a) Bacteria (Sphaerotilus natans)

Brünlisbach

50 m

(b) Tubificides

700 m

1,8

Mettma

50 m *probe position. Cloudy water, simplest forms in drop picture. Decomposition zone for substances. Bacteria dominate.*

◄ 700 m, *cloudy water. Drop picture same as at 50 m, discshaped. The oxygen content in the decomposition zone is smallest. Bacteria and worms dominate.*

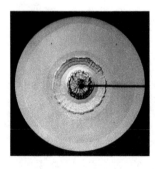

1,800 m. *Disappearance of bacterial flakes. Beginning structuring and contraction of the disc of the drop picture. A large part of the substances have been degraded. Increased appearance of wormlike insect larvae .* ➤

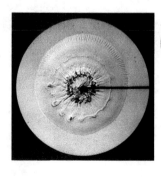

The wastewater discharge causes a complete change in the habitat and the living community there:

— the streambed becomes covered with a furry bacterial growth: *Sphaerotilus natans* (see figure *a,* opposite page)
— a massive appearance of a few types of organisms
— living on the bed we find tubificides *(b)* and non-biting midges *(d)*

From 1,800 metres onwards black-fly larvae *(e)* dominate the scene. During the further development down the section ever more species of sessile algae appear and no single form dominates any longer. Between 2,000 m and 3,000 m ever more sessile algae species inhabit the bed. With the algae, herbivores — grazing micro-organisms such as gastropods *(f)* and mayflies *(k)* — appear again.

By the end of the self-purifying section, the plant and animal communities appear again in the balanced ecology which previously existed in the undisturbed stream.

The Mettma after 6,000 m

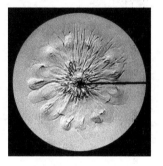

8,000 m. *Balance in the drop picture, in the chemical processes and in the plant and animal communities. The substances are mineralized. The stream has cleaned itself.*

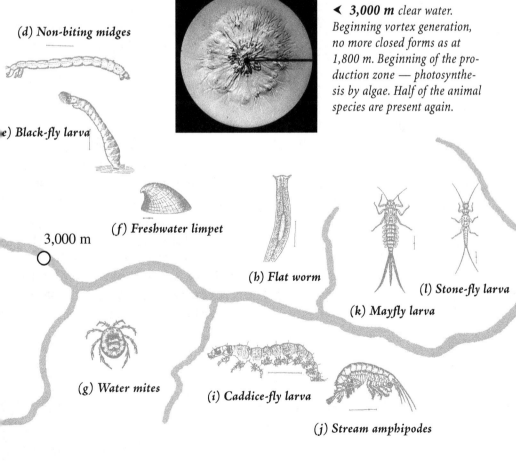

◄ **3,000 m** *clear water. Beginning vortex generation, no more closed forms as at 1,800 m. Beginning of the production zone — photosynthesis by algae. Half of the animal species are present again.*

(d) Non-biting midges

e) Black-fly larva

(f) Freshwater limpet

3,000 m

(h) Flat worm

(l) Stone-fly larva

(k) Mayfly larva

8,000 m

(g) Water mites

(i) Caddice-fly larva

(j) Stream amphipodes

The waterway has regenerated after 8,000 m. It has become a mountain stream again with clear water, offering an environment for differentiated life.

75

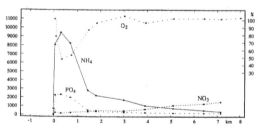

▲ *Ammonium (NH$_4$) nitrate (NO$_3$) and phosphate (PO$_4$) concentration in µg/l, as well as the oxygen saturation in %.*

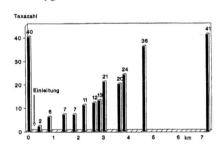

THE FAUNA CORRESPOND TO THE DROP PICTURES

Organic substances and metabolic processes dominate the polluted part of the stream. Organisms with a simple body structure appear: wormlike metabolically active animals with a simple organization. In the drop pictures the flow behaviour is clearly impeded in its formative ability — discs and ring-shaped forms dominate.

In the clear water of the natural stream we find organisms with a differentiated body form (head, thorax, abdomen clearly distinct) and differentiated sense activity. In the drop picture experiment the mobility of the water leads to a multiformed form with open shapes. Vortices and line structures interpenetrate one another.

◀ *Number of species (taxa) of animals on the stone substrate (after Schreiber, I. [6]) during the course of the self-purifying section.*

Handling of drinking water

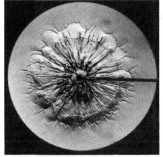

Effect of container material
Plastic — Glass

Distilled water from a synthetic bottle

Distilled water from a glass bottle

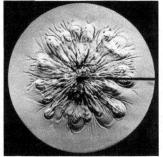

Effect of the piping material
Plastic — High-grade steel

Drinking water after being in a pipe made of PP-R (polypropylene-Random-copolymerisate) for 20 hours at 20°C

Drinking water after being in a high-grade steel pipe (V4A) for 20 hours at 20°C

Stagnation in the piping under pressure and heating
Stale — Fresh

City water after many hours of stagnation in the piping

... from the same tap after rinsing the piping

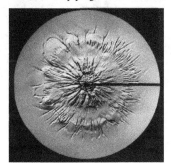

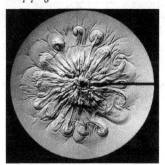

Purifying measures
Kitchen paper — Laboratory paper

Experimental dish wiped dry with a kitchen paper-towel and then rinsed

Experimental dish wiped dry with laboratory filter paper (MN 710) and then rinsed

The effect of surfactants

Surfactants are substances which, among other things, reduce surface tension, like detergents. Even at a concentration of 1 part per million (ppm) a tenside (the surface-active organic substance in detergents) can change the flow during a drop picture experiment.

Experiment with pure water. Surface tension 72.66 milliNewtons per metre (mN/m). Vortical, multiformed with a sharply bounded kernel region.

5th	*9th*	*20th*	*37th*

Experiments with a diluted tenside (TPS: Tetrapropylenebenzosulphonate)

1:1,000,000 dilution — 72.64 mN/m surface tension. The kernel region is no longer sharply bounded.

1:500,000 dilution — 72.60 mN/m surface tension. Vortex generation has weakened, the radial line structures become amplified.

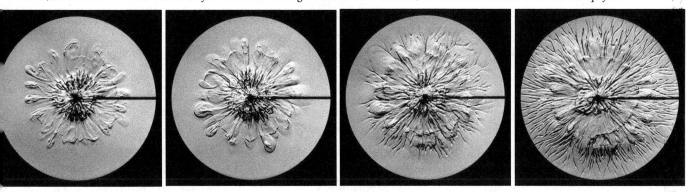

1:250,000 dilution — 72.49 mN/m surface tension. The picture contracts, in second photograph the radial line structures dominate the picture.

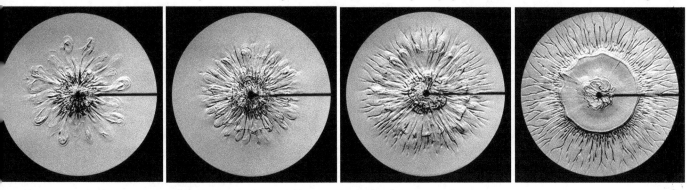

1:125,000 dilution — 72.27 mN/m surface tension. The forms in the middle region disappear.

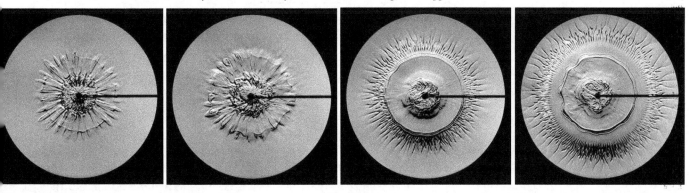

1:50,000 dilution — 71.48 mN/m surface tension. The spreading drop fluid generates concentric rings only.

5. Basic Research: Drop Phenomena

The aim of basic research is to try to understand the natural world. This has consequences for the way in which we conduct our lives and how we care for the earth, the foundation of our existence. Researching flow processes contributes specifically to our understanding of the drop picture method, as well as enhancing our general understanding of water.

When we observe such processes, we find we are also studying universal processes which can shed a more general light. We see, for example, how form arises out of motion, how rhythm springs forth, or what 'fluid' really means, and we also realize how far we still are from making our own thinking more mobile.

In this chapter we will be concerned with drops and their impact on a water surface. The phenomena we describe here are aspects of water behaviour, like those described above. We have, however, placed them at the end because they occur so quickly that little can be seen by direct observation of the unaided eye. Here we have a new challenge which calls on our imagination to recreate the process by means of single photographs.

In our research on flow processes during a drop picture series, we find the following are the most fruitful research areas: the drop formation on the end of a needle, the change of form of the falling drop, the impact of the drop on the liquid surface and the subsequent vortex generation underwater.

Drop formation

The most universal of all forms is the sphere: that form to which a water drop always tends. It has no top, no bottom, no left, no right. It has a relationship only to its own centre and to the periphery. It is a form which water only really takes on when tiny droplets float in the atmosphere (see pp. 12f). As a drop grows its relationship to gravity increases, and it begins to fall. Now its surface becomes flattened by the air through which it travels: a top and a bottom are formed, and the flowing air makes it oscillate as it falls. Out of the universal, something individual arises: the timeless hovering becomes a specific falling and oscillating. When the drop becomes too large, it can no longer keep itself together and splits into smaller droplets. This is a process of detachment, as during cell-division or during metabolism when the vesicles of the cell divide.

This process of detachment also happens at the drop-picture needle. First, a small round drop generates which slowly becomes more pear-shaped as it grows longer and larger. This outward form of the drop is the expression of the forces acting on it. When the strength of gravity's force on the drop nears that of the surface tension, the drop elongates as it starts to detach. This is an important moment, in that the tiniest vibration — even that due to a sound — can either accelerate or delay the detachment. In the first case the drop will be smaller while in the second, it grows bigger. As it detaches, it begins to oscillate and a tiny ring-shaped wave migrates over the drop. This wave pushes a multitude of waves in front of it. These waves are called capillary waves because they are damped by surface tension. It is these forces which makes water rise through thin tubes (capillaries), the forces which play a part in causing the sap to rise in plants.

When such a wave arrives on the underside of the drop it concentrates to a point causing this region to protrude substantially. In this way these tiny waves can be recognized by the relative pointedness or flatness of the upper or lower surface.

These strong deformations cause its optical reflective and refractive properties to change, so that the waves and oscillations can be studied from their effect on the light. The drop detachment is like the moment of birth leading to a series of events that follow a strict pattern.

One can imagine that a drop form, flattened at the bottom, impinges in a different fashion than when it is pointed. Indeed, the form as well as the motion of the drop at the instant of impact on to the water layer play an important role in the drop picture.

As far as the drop picture method is concerned, the drop is the trigger which brings the sample fluid into motion: this impulse is differentiated. It is therefore important for us to know — and be able to repeat — the exact initial conditions. Page 60 shows some drop pictures of experiments carried out at various drop heights and with various drop forms at the moment of impact.

A drop is not a drop

Drops generated from condensation initially form a sphere. Drops which separate from a water jet or a needle, for example, begin to oscillate and change their form while falling.

A drop forms at the end of a needle (outer diameter of 0.8 mm). The surface tension maintains the spherical form as long as gravity does not draw it out. The needle tip is visible above each drop.

Because of the separation of the drop from the needle, a ring-shaped wave forms which creates further fine capillary waves as it moves. These waves run about the drop. This leads to a complicated oscillating form. The lower end of the drop shows a sharpening and a flattening, due to the superposition of the many waves.

What the eye doesn't see

Strong light reflections may be caused by the raised fine capillary waves on the falling drop. These cause the drop to glint (see arrows). The oscillating drops light up like a string of pearls.

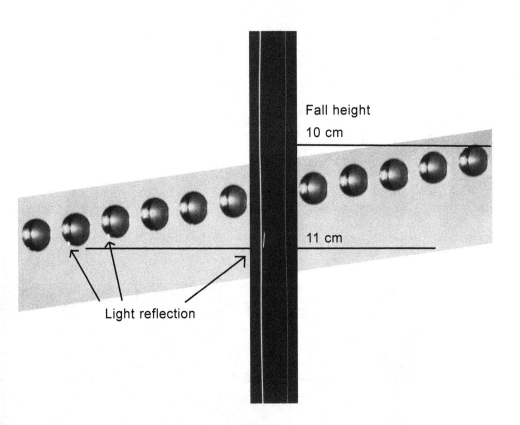

Fall height
10 cm

11 cm

Light reflection

The stripes to the right and the left of the photograph are light reflections on and within the falling drop. The short streak marks a brief interval during a particular oscillation phase.

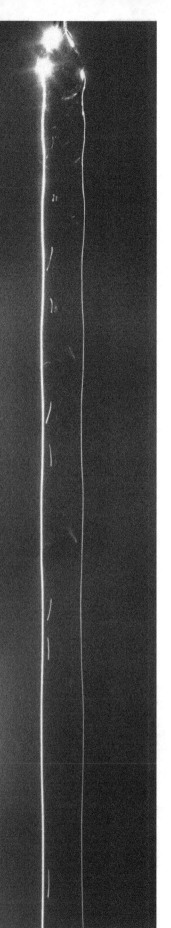

◄ *The falling oscillating drop has left a path of light behind it —*
flashes at a particular point witness to the drop's changing form

A hanging drop in the light against a dark back-
ground. To the right the light is reflected on the
surface. To the left, the reflection is more from within.

The intermittent path of drops splashing from a stream shows how their oscillation is
caught by the flashes of the sun

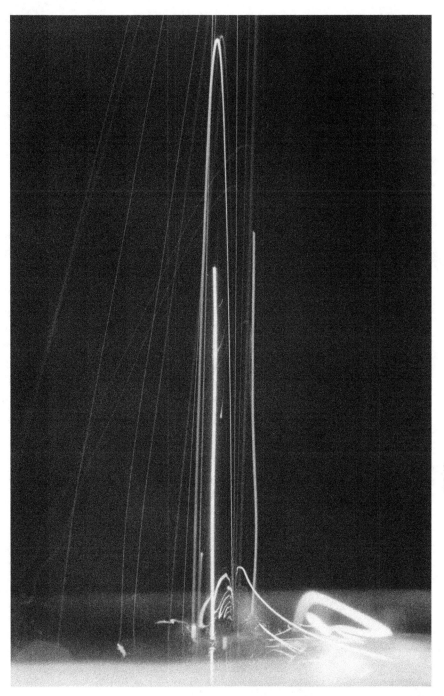

A drop released into water from a height of 10 cm and illuminated from a slanted angle. The time exposure shows how many droplets are splashed upwards, each of whose path is traced by a line of light.

The immersion of a water drop

When a drop falls into standing water it is not immersed immediately, but rather its impact sets up a complicated interaction of motion and forms during the displacement of the water. A hemispherical crater forms, during which the drop spreads as a thin layer. The displaced water simultaneously shoots upwards past the edge of the crater and creates a lamella-like crown, which atomizes into fine water jets which in turn break into many individual droplets. Once crater and crown have reached their maximum extension, they begin to take on a new form due to the change in direction of their motion. The water of the crown moving outwards generates a ring-wave, while the water moving inwards flows into the crater. It shoots from all sides towards the centre, leading to an impinging region from which a water jet moves upwards. This jet may later divide into droplets. Droplets and jet sink back into the water, generate vortex rings and only now can the drop fluid unite with the standing water.

Drop and water surface

The entire impinging process goes through several phases in which repeatedly a motion leads to a form and vice versa. Initially there are two polar opposites: the falling drop, with its form reduced to a point, and the standing water with its extensive surface.

PICTURE SEQUENCE: From a height of 60 cm, a blue dyed drop falls into red dyed water. The drop fluid spreads as a thin layer in a crater. From the very first moment, at the edge of the crown, it is mixed with lower-lying fluid and slung outwards as droplets.

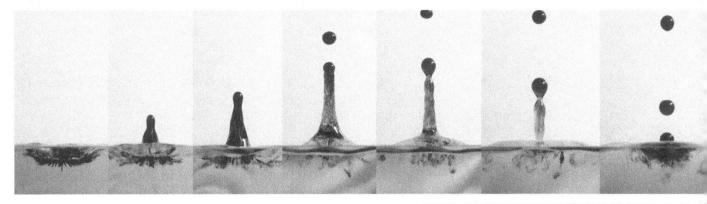

The impact

Water meets water. The drop is just as 'hard' for the water surface as the latter is for the drop. It cannot penetrate but its motion must first be halted while the standing water is displaced. This creates such a violent motion that the water is tossed to the side like dust.

Generation of crater and crown

During the transition from the first violent contact to the slower flow, the drop and the surface are both deformed. The increasing deformation is accompanied by an outward moving displacement of the water from the impact crater. The motion then changes to a decelerating displacement flow which again rises into the crown until it comes to rest there. The momentum of the drop has now passed completely into the form and, in the same manner, the drop fluid has spread out superficially as a thin layer in the crater, part of which has flowed from the crown as a splash. A falling motion of the drop has become completely dissipated throughout a newly created surface form.

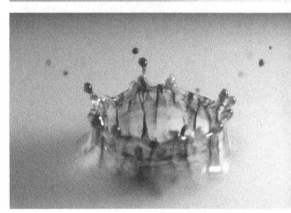

Generation of a new reverse impulse

This form of the water surface is not stable: a wave ring spreads out while generating capillary waves in front of it. A reversal of the motion takes place within the crown and the water flows together in the crater. The drop fluid dissipated throughout the crater, is gathered together. A new motion has formed from the unstable form.

Impinging point and jet generation

The water masses coming from the periphery flow together towards a point at the centre and impinge with a force sufficient to project an amount of water upwards — far more than the original quantity

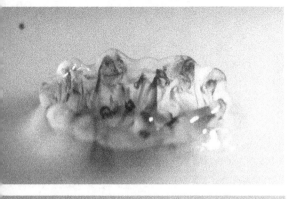

of drop water which fell. A water column, or jet, rises, often dividing into hovering drops. Polar to the hollow crater form we find the filled-out water column.

Immersion and mixing

A second moment of reversal of the motion occurs when the jet begins to fall. The water column sinks back and the detached drops can now immerse, since they move slower. The fluid is drawn out into a large surface by an underwater vortex ring motion which dissipates as it comes to rest (pp. 32–7). Only now has the drop been completely subsumed by the standing water.

Spreading of the water

If we follow the process in the mind's eye, we know how the entire water level rises when we add water. The motion only ends when the different water levels have completely levelled out. We find the flow process of this leveling out in the river forms, in treelike structures and in meanders.

Shattering – deformation – flow

The first moment of the impact reminds one of an event where solid bodies collide with each other and shatter. But the more we follow this process as it develops, the more it becomes an expression of fluids. The deformation is a first sign of fluid motion. We find crater generation when meteors hit the earth, the moon, and so on. Even with thick mud we find the beginnings of jet generation, and the more fluid the medium, the further the process develops, until we arrive at the myriad forms in pure water.

About motion in space

If we consider what happens to the drop, we may also see it as a way into the inner dimensions of space. First, the linear fall, then the spreading out into the surface (crater), then the filled form of the

PICTURE SEQUENCE: The crown becomes a wave. The remaining drop fluid shoots back towards the center of the crater. Virtually the entire remaining drop fluid is slung out of the water on the end of a rising jet, before it finally returns, and mixes with the hitherto still water.

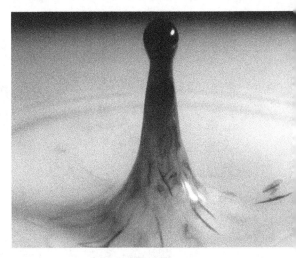

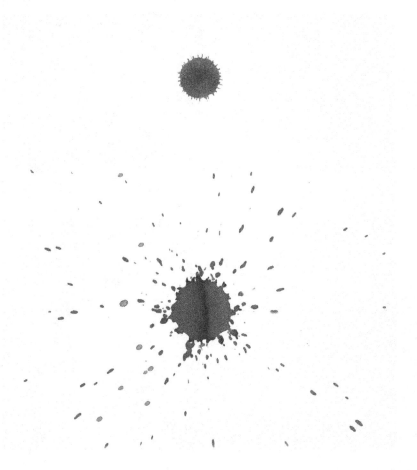

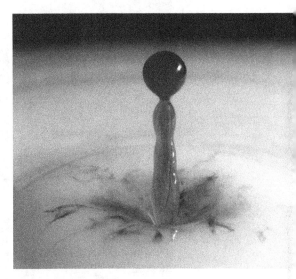

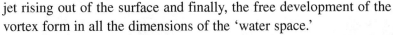

▲ *A drop of ink on paper, below: five drops. The first drop does not splash. The second drop falling onto the first causes the splash.*

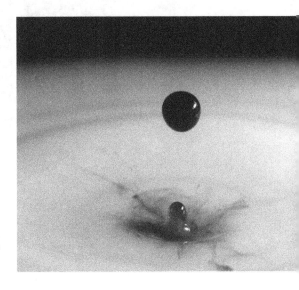

jet rising out of the surface and finally, the free development of the vortex form in all the dimensions of the 'water space.'

In the drop picture experiment many drops fall from a height of 10 cm, that is, with not too great a speed, into a shallow fluid layer. Instead of generating a crown, a ring-wall forms in which the region of generation of the drop picture vortices may be seen. The crater is bounded by the floor of the shallow dish, preventing a jet rising. The drop water remains as the kernel zone at the centre so that it can be further driven to the periphery by the next drop, where it forms a part of the schlieren of the next vortex.

A universal process

All development processes follow universal laws. Thus through studying this immersion process we discover aspects that occur equally in the most varied forms of life, growth and decay.

LEFT: A drop falls onto a glass plate, spreads and again draws together without splashing.

MIDDLE: A drop falls from 10 cm height into deep water. A crater, surrounded by a ring-shaped wave, is generated. A jet then forms, which throws the finest droplets high into the air (see p. 85).

RIGHT: A high velocity drop falls from a height of four metres into water. A large crater forms and a crown shoots upwards, enclosing an air bubble.

The immersion in the drop picture experiments

The first moments — invisible with the naked eye — after the impact of the drop on to the test sample in the drop picture experiment. Filmed with a video camera with an exposure time of 1/10 000 of a second.

1. A crater and a ring wave form. The ring wave pushes an array of capillary waves in front of it.
2. The wave has moved out, allowing the forming drop picture to become visible.
3. Garlands and vortices, here just distinguishable in their first phases, are generated from the ring wave.
4. While the vortices are generated, the waves, reflected from the edge of the dish, turn in again towards the centre, the smallest first.
5. The larger reflected waves move in towards the centre.
6. It takes less than a second — the second on the time counter has not changed — for the drop picture to form. It can be photographed after 1.5 to 2 seconds by which time all wave motion has subsided.

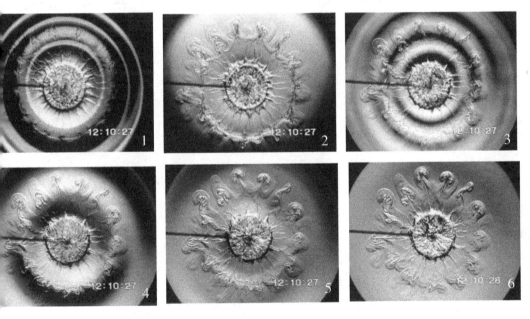

The stage shown in the top left picture (1) is now shown in more detail, viewed at an angle. One sees how the crater with the ring shaped bulge forms, pushing capillary waves in front of it. This series took less than a tenth of a second. The waves of the last photos are doubly visible due to reflection on the bottom of the dish. ➤

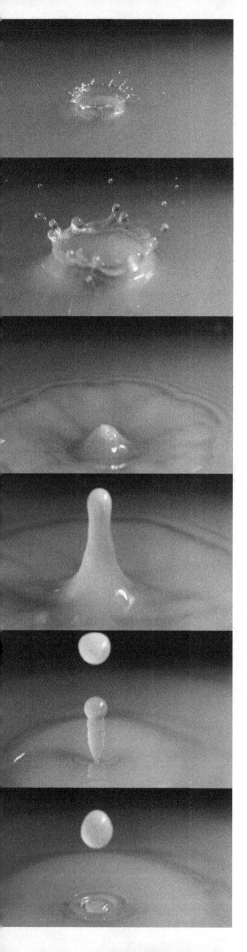

A drop falls into water

A *pulse* changes into a rhythmic series of *forms,* in which each is displaced by the next, while the last, as they come to rest, again lose themselves.

Falling and carrying momentum, the drop approaches the surface in that most universal of all forms — the sphere:

1. Like a hard solid body, the drop hits the water surface. The water is slung sideways and atomized to 'dust.'
2. A crown forms and atomizes upwards as droplets. A crater forms due to the spreading of the drop fluid as a fine film. The impulse has lost itself in the form. The new form is not stable and a counter-motion begins.
3. Water flows from all sides towards the center of the crater — a rising motion begins. Water shooting back towards the center impinges upward.
4. A jet forms. Ring-waves spread out.
5. Drops detach from the jet.
6. The drops and jet sink back and immerse themselves in the water, where they generate vortices.

Conclusion

Water presents us with a plethora of phenomena, each filling us with wonder. But what is the importance of researching these phenomena? In ancient times people seem to have been given the wisdom of being able to work with water in the right way. Ancient cultures developed sophisticated water engineering techniques. However, there are also many historical examples in which a thoughtless relationship with water led to plague and poverty. The earlier wisdom protecting humanity against plague and disease, now has to be consciously understood and used. Every one of us must be a responsible person as far as water usage is concerned. However, we can only act responsibly once we understand the interrelationships within nature.

Nature and life develop in a cyclical and rhythmical fashion. Yet we human beings often do not consider what goes on before and after our activities. With our technical know-how everything seems possible, yet we continue destroying life. This cannot continue much beyond the present. Our future must be one in which we look at life on a larger time-scale. This means that we must comprehend the cyclical workings of nature, demanding nothing less than comprehending the earth as a living organism.

Our water research is not an end in itself, simply to enjoy the beauty of water phenomena, but rather to learn to understand what, in nature, acts as beautiful and healing, in order to incorporate health and beauty into our fashioning of the natural world. From water we can learn to find the middle way between extremes, to mediate between opposites.

One could quote many poets and philosophers who see beauty as being the highest expression of the unity between mind and matter. Beauty in human achievement may become a pointer towards health — in its broadest sense. It is in this sense, that the incorporation of beauty in research can become something near to life and by no means impractical.

In trying to point something out which is not generally accepted, one cannot expect immediate approval. Convinced of the purpose of our work, we hope to encourage readers of this book to live with these ideas, and to feel a need to develop them further.

Founding members and scientists in 1960 (from left): Dr Georg Unger (1909–99), George Adams (1894–1963) and Theodor Schwenk (1910–86)

Appendix

The Institute of Flow Sciences, Herrischried

The focal area of research at the Institute of Flow Sciences (Institut für Strömungswissenschaften) is the question of what quality is — what makes good and enlivening water in all its connections with life. The aim is to understand water as a mediator for life, and the work is based on scientific principles necessary for developing a responsible attitude to water.

Situated in Herrischried, in the Black Forest, Germany, the Institute was founded in 1961 by Theodor Schwenk, a fluid mechanical engineer, who was its first director. In the 1950s, Theodor Schwenk and the mathematicians Georg Unger and George Adams had begun working together. Investigating the dimension of flow, Theodor Schwenk soon realized the importance of projective geometry, a method particularly suited for investigating organic forms. The mathematical geometrical approach to the treatment of the motion of water could not be continued after the early death of George Adams in 1963.

Through Alexandre Leroi, Adams, Schwenk and Unger obtained generous funding from Hanns Voith to found the Institute, and the Weleda Company in Schwäbisch Gmünd, Germany, freed their longstanding co-worker Schwenk for this new task. The Weleda Company supported the Institute for many years. Schwenk, together with Helga Brasch (physician and eurythmist), were the primary developers of the Institute.

In the 1960s, the Institute became known through Theodor Schwenk's book *Sensitive Chaos,* in which he describes his water research. He developed the drop picture method, an approach designed to show the more subtle qualities of water as a flow picture. The research method focuses on water quality, and complements analytical methods with a more holistic concept of quality. The approach is image-generating rather than analytic, and is based on Goethean science principles. It brings the formative and renewing activity of water into images.

Theodor Schwenk

Dr Alexandre Leroi (1906–68) Doctor, co-founder and formerly member of the Board

Dr. Hanns Voith (1885–1971) Mechanical Engineer, factory owner, co-founder, chairman and sponsor

John Wilkes, who was working in creative design at the Institute in the early years, developed his first Virbela Flowforms here in 1970. These are cascades of specially formed basins through which water flows, causing it to move rhythmically.

Since the 1980s the authors of this book have continued the Institute's work. The drop picture method has been further developed in order to understand better its fundamentals as well as its predictive powers. It remains the central tool to this day. We are interested in practical applications and especially in developing principles of what good drinking water can be. Past research of the Institute on drinking water, groundwater, spring water and river water show that the particular mobility of each kind of water leads to different images. But we are also interested in understanding technical influences on the mobility of various kinds of water — influences such as industrial processing, water treatment, and so on. We also research basic questions, for example, concerning rhythm-research. We do not, however, test products or write reports on water-processing products.

The Institute in 2005

Capillary waves

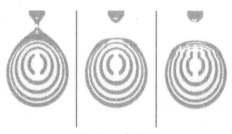

Drop oscillations

Drop picture formation

Our research work is published in our own journal *Sensibles Wasser*. In addition to a collection of Theodor Schwenk's lectures concerning ethical questions about water (Vol. 1) the journal contains specialist contributions on morphological terminology and instructions on evaluating drop pictures (Vol. 2); and groundwater investigations over many years aimed at characterizing an ideal of good quality water (Vol. 3) as well as comparative investigations on the biological self-purifying section of a brook (Vol. 4). Here the drop picture method shows an integrated snapshot of the water condition in the same way as the study of biological population shows a holistic but temporally integrated picture. Volume 5 gives a comprehensive technical description of the drop picture method.

Research in the past few years has continued with refining the evaluation to give more differentiated and subtle results. Interest has grown in the drop picture method as a means of giving a qualitative evaluation of water. Specific research has been done on the effect of container materials on water, for instance glass or PET (polyethylene terephthalate) bottles, and on the effect of physical water treatment with movement. Tests on the methods of stirring biodynamic compost preparations have also been undertaken.

At the core of this research is the striving to understand the nature of water, and to impart this to others. Additionally we have developed simple experiments to demonstrate the special quality of water (for instance the wave experiments on pages 38f).

Our work also finds an ever-increasing interest within specialist circles, leading to a great demand for publications, tours and seminars.

At present the Institute is staffed by four scientists. As an independent Institute, we battle with financial problems, and our activities are largely financed by gifts and donations, supplemented by fees from research contracts and revenue from publications. We are making great efforts to maintain and extend the circle of sponsors, who are always welcome.

We are lucky to have committed members in the board of trustees, all experts in their fields who, together with the organization's committee members and scientists, advise the various decision-making bodies. The commitment of members and co-workers arises from their realization that the roots of problems concerning water lie not in a lack of technology, but in a lack of inner attention to water itself, in the indifference to this mediator of life. Neither moralizing appeals nor apparatus will change anything; rather awakening and educating our ability to pay attention to the activities of water and its meaning for life.

The Institute of Flow Sciences has been visited by thousands of people many of whom became interested and excited by their experiences there. We aim to follow up on and extend this work. When seen in this manner the intended 'products' of the Institute are not primarily technical know-how, but the generating of awareness and encouraging a feel for water on the basis of modern anthroposophically oriented scientific research. Appointments for visits may be obtained on request from the office.

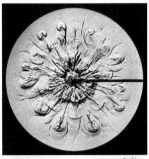

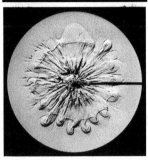

▲ *Pure spring water and tap water of an industrial city*

Institut für Strömungswissenschaften
Stutzhofweg 11
79737 Herrischried, Germany
Tel. +49-7764-9333 0
Fax +49-7764-9333 22
E-mail: sekretariat@stroemungsinstitut.de
www.stroemungsinstitut.de

▼ *Water before and after contact with synthetic substances*

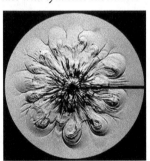

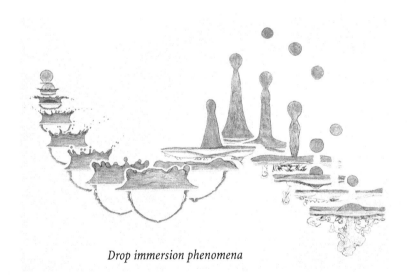

Drop immersion phenomena

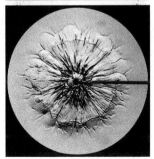

We honour all holy water
Which stills the thirst of the earth, all the holy water
And all the plants, which the Creator made.
For all are holy!

We honour the Waters of life,
And all that is water on earth.
Still, flowing and springing water
The ever flowing springs
Blessed raindrops
We devote ourselves to the holy waters
Which have created the law.

Heavenly Father! And you, Angel of Water!
We thank You and praise Your Name.

A flood of love springs forth
From the hidden places beneath the earth:
The brotherhood is blessed forever
By the holy Water of Life.

From the *Writings of the Essenes*

Glossary

Capillary waves. Surface waves are distinguished according to the primary force tending to restore them to an undisturbed surface. Both gravity and surface tension (capillarity) act in the above sense and are superposed on each other. For longer wavelengths the force of gravity predominates. For very short wavelengths (smaller than 17.2 mm) the force of surface tension predominates, and these waves are called capillary waves.

Density of a substance is the ratio of its mass to its volume. If the mass is measured in grams and the volume in cubic centimetres then the unit of density is g/cm^3 (which is the same as kg/litre). For a particular volume of a substance, the denser it is, the heavier it will be, and the more inertia it will tend to have.

Harmonic flow is the sensitive transition range between laminar and turbulent flow.

Parameters are physical characteristic or feature which can vary. The height of fall or drop-size for the drop picture experiment, are two examples. Similarly, the density, viscosity or the surface tension of the sample are three more parameters.

Schlieren are the density gradients or boundaries within transparent fluids. They appear as the air flickering over a hot surface or flickering when salt is dissolved in water.

Surface tension is the interfacial tension between the surface of two different substances (for instance, liquid/gas, liquid/liquid or liquid/solid) which acts to pull the interface together to make it as small as possible. It acts against the tendency of liquids to wet solids; like a uniformly stretched extensible and very thin skin.

Surfactants are surface active agents, like detergents.

Shearing. Within the fluid differences in the velocity between neighbouring regions lead to an ever increasing separation of former neighbours. A compact region of dyed fluid is drawn out into an ever longer thinner region.

Viscosity of a liquid is its tendency to resist every motion. Honey, glycerine and thick oils are examples of particularly viscous fluids.

Vortical. The more or less stronger development and shape of the rosette-formed vortices in the drop picture. The term 'vortical' is used purely morphologically for describing figures and is not the same as the technically used word 'vorticity' (local spin of the fluid).

Photograph acknowledgments

Page

15 Rising air bubble: *Physik in unserer Zeit*, 1981, No. 5, p. 158

16 Waves from wind: Markus Wilkens

19 Vortex funnel: Theodor Schwenk

23 Meander in Stone: Herbert Glatz

29–31 Vortex streets: Theodor Schwenk

29 Vortex generation: F. Homann

32 Vortex funnel: Herbert Völkle

74f Animal drawings: from Wolfgang Engelhardt, *Was lebt in Tümpel Bach und Weiher,* with kind permission of Franck-Kosmos-Verlag, Stuttgart

74f Landscapes from Mettma: Heinz-Michael Peter

90 Drop impact onto a glass plate: Verein für Krebsforschung, Arlesheim

94–97 Archive of the Institute of Flow Sciences

All drop picture photographs from the Institute of Flow Sciences, Herrischried

All other photographs and drawings: Andreas Wilkens

References

1. *Sensibles Wasser.* 1/85, Herrischried

2. Schwenk, Theodor. 1967. *Bewegungsformen des Wassers. Nachweis feiner Qualitätsunterschiede mit der Tropfenbildmethode.* Stuttgart. Freies Geistesleben.

3. Matthijsen et al. 1994. 'De druppelbeeldmethode als potentiele integrale somparametrische indicator voor waterkwaliteit.' *H twee O* (Rijswijk) 27. 22.644–47, 660.

4. Selawry, Alla. 1987. *Ehrenfried Pfeiffer, Pionier spiritualler Forschung und Praxis.* Dornach. Goetheanum. p. 90.

5. Steiner, Rudolf. 1998. *From Elephants to Einstein.* London, Rudolf Steiner Press. Lecture of February 9, 1924.

6. Schreiber, I. (1975) 'Biologische Gewässerbeurteilung der Mettma anhand des Makrozoobenthos: Methodenvergleich.' *Arch. Hydrobiol. Suppl.* 47, 432–457

Bibliography

1. BOOKS AND ARTICLES FROM THE INSTITUTE

Jacobi, Michael. 2004. 'Auswertungsmethodik beim Vergleich von Herstellungsvarianten biologisch-dynamischer Rührpräparate mit der Tropfbildmethode.' *Elemente der Naturwissenschaft.* 81.95–102.

Picariello, Christine. 2004. 'Schritte der Auswertung von Tropfbildern.' *Elemente der Naturwissenschaft.* 81.91–94.

Schwenk, Theodor. 1965/1996. *Sensitive Chaos.* London, Rudolf Steiner Press.

Schwenk, Theodor & Schwenk, Wolfram. 1990. *Water, the Element of Life.* New York. Anthroposophic Press (Nine essays by Theodor Schwenk and four essays by Wolfram Schwenk).

Schwenk, Wolfram. (ed.) 2007. *The Hidden Qualities of Water.* Edinburgh, Floris Books.

Schwenk, Wolfram. 1995. 'The mobility of water as an aspect of quality and its visualization by means of the drop picture method.' Prague. *Ziva Voda* 95.20–27 & 85–95.

—. 'Water as an open system.' In H. Dreiseitl et al. (eds.) 2001. *Waterscapes. Planning, Building and Designing with Water.* 106f. Basel, Berlin, Boston. Birkhäuser.

—. 2001. 'Wasser, das universelle Lebenselement.' *Elemente der Naturwissenschaft.* 74.1.8–25.

—. 2004. 'Die Rolle der Stoffe bei den Gestaltungsprozessen in der Natur und bei den bildschaffenden Methoden.' *Elemente der Naturwissenschaft.* 80.108–13.

Wilkens, Andreas. 2004. 'Strömungsvorgänge beim Tropfbildversuch und Beziehungen zwischen Probe, Strömungsprozess und Bild.' *Elemente der Naturwissenschaft.* 81.5–22.

Wilkens, Andreas. 2005. 'Betrachtungsebenen - von der Morphologie der Tropfbilder bis zum Lebenszusammenhang' *Elemente der Naturwissenschaft.* 83.33–51.

Wilkens, Andreas. 2006. 'Vergleichende Tropfbildversuchsreihen als Instrument der Urteilsbildung' *Elemente der Naturwissenschaft.* 85.40–56.

2. Journal from the Institute of Flow Sciences: *Sensibles Wasser* (ISSN 0178-7047)

1 (1985). Schwenk, Theodor. *Das Wasser, Herausforderung an das moderne Bewußtsein* [An ethical approach to modern thinking about water].

2 (1993). Jahnke, Dittmar. *Morphologische Typisierung von Tropfenbildversuchen und Tropfenbildern. Morphologische Unterscheidungsmerkmale für die Auswertung von Wasserqualitäts-Untersuchungen mit der Tropfenbildmethode* [Morphological description and classification of drop pictures].

3 (1994). Jahnke, Dittmar. *Langjährige Grundwasser-Untersuchungen mit der Tropfenbildmethode. Beitrag mit der Tropfenbildmethode zum Trinkwasser-Qualitätsleitbild nach DIN 2000. Gegenüberstellung von hydrochemischen Analysedaten und Versuchsergebnissen mit der Tropfenbildmethode* [Comparative studies with the Drop Picture Method and chemical analyses on water from different ground waters].

4 (1994). Peter, Heinz-Michael. *Das Strömungsverhalten des Wassers in der biologischen Selbstreinigungsstrecke des Schwarzwaldbaches Mettma. Untersuchungen mit der Tropfenbildmethode im Vergleich mit biologischen, chemischen und physikalischen Parametern* [Comparative studies on the biological self-purification of a polluted stream by means of biological, chemical and drop picture phenomena].

5 (2000). Wilkens, Andreas, Michael Jacobi & Wolfram Schwenk: *Die Versuchstechnik der Tropfenbildmethode — Dokumentation und Anleitung* [Technical set up and experimental instructions for the Drop Picture Method].

6 (2001). Schwenk, Wolfram (ed.). *Schritte zur positiven Charakterisierung des Wassers als Lebensvermittler. Ausgewählte Texte aus 40 Jahren Wasserforschung mit der Tropfenbildmethode* [A selection of publications of 40 years of water research, mainly with the drop picture method, towards characterizing water as life giving element].

7 (2001). Metzler, Franz (ed.). *Wasser verstehen — Zeichen setzen. 40 Jahre Institut für Strömungswissenschaften 1961–2001* [Chronicle and lectures published for the fortieth anniversary of the Institute of Flow Sciences].

3. Other English language books of interest or relevance

Adams, George & Olive Whicher. 1980. *The Plant between Sun and Earth.* London. Rudolf Steiner Press.

Alexandersson, Olof. 2002. *Living Water.* Dublin. Gill & Macmillan.

Bartholomew, Alick. 2003. *Hidden Nature.* Edinburgh. Floris.

—. 2010. *The Story of Water.* Edinburgh. Floris.

Coats, Callum. 2001. *Living Energies*. Dublin. Gill & Macmillan.

Consigli, Paolo. 2008. *Water, Pure and Simple, The Infinite Wisdom of an Extraordinary Molecule*. London, Watkins.

Dreiseitl, H. *et al.* (eds). 2001. *Waterscapes. Planning, Building, and Designing with Water*. Basel, Berlin, Boston. Birkhäuser.

Edwards, Lawrence. 2006. *The Vortex of Life*. Edinburgh. Floris.

—. 2003. *Projective Geometry*. Edinburgh. Floris.

Probstein, Ronald F. 1989. *Physicochemical Hydrodynamics*. Boston, London. Butterworth.

Jenny, Hans. 2007. *Cymatics: A Study of Wave Phenomena and Vibration*. Macromedia Press.

Lauterwasser, Alexander. 2007. *Water Sound Images: The Creative Music of the Universe*. Macromedia Press.

Somero, G.N., Osmond, C.B., Bolis, C.N. (eds). 1992. *Water and Life*. Springer.

Whicher, Olive. 1985. *Projective Geometry*. London. Rudolf Steiner Press.

Wilkes, John. 2003. *Flowforms: the Rhythmic Power of Water*. Edinburgh. Floris.

Index

Printed in the USA
CPSIA information can be obtained
at www.ICGtesting.com
JSHW052026301024
72691JS00006B/32

9 781782 505068